Build The Wall

including

Climb The Wall

2 sets of photocopiable worksheets to encourage
and aid with the learning of times tables.

The sheets are mainly for pupils at primary, middle
or lower secondary/high schools.

by

Tony Colledge

ISBN-13: 978-1503277762
ISBN-10: 1503277763

Digital Master: 2.1

For comments, feedback or suggestions you can contact the author via email at

beatthewall@yahoo.com

Other books published by the author include

Beat The Wall

The Dividing Wall

Pascal's Triangle

Other books to be published in 2015 include

The Fraction Wall

Beat the Deciwall

About the Author

Tony Colledge was educated at Bishopshalt Grammar School in Hillingdon, London. He gained a BSc in Civil Engineering at Swansea University in 1976 and then the following year he trained as a teacher at Gypsy Hill College, Kingston in London, where he received his PGCE.

He spent 37 years in England teaching maths across the whole ability range to pupils aged from 8 to 16. In 1992 he wrote his first book about looking for patterns in Pascal's Triangle.

When he retired in July 2014, he decided to adapt and publish some of his most successful resources in the hope that other teachers, parents and pupils might find them useful.

About the Books

Beat The Wall resources have been published as a set of books each containing photocopiable worksheets based on a particular area of the Numeracy Curriculum. The original worksheets were successfully used to raise pupil achievement, not only in his own classroom but also by other maths teachers working in the same school. They have also been used in lessons observed by Ofsted inspectors and members of the senior management team on many occasions. The lesson gradings in which these resources were used, always resulted in a classification of either good or excellent/outstanding.

Tony Colledge created the first Wall sheets in 1995 whilst teaching several low ability maths classes. He realised, like many other teachers, that the difficulty those pupils had with learning and recalling tables was like a barrier, or *wall*, which hindered their progress in several areas of the curriculum. Of the many resources he wrote at the time, the Wall sheets were the most successful at improving scores. In fact, they were often requested in lessons by the pupils themselves. Weak and able children alike recognised the progress they were making and enjoyed the challenge of trying to beat their previous scores and times.

Since then, he has continued to develop, trial and improve a range of similar sheets and he is convinced that if these resources are used regularly, and as advised, then all pupils will improve their knowledge and understanding of numeracy in the areas covered by the worksheets.

Using the worksheets

The worksheets are **not** designed to be used as an initial teaching resource.

They are more valuable if used as a measure of pupil understanding, application and recall of numeracy facts once they have been introduced to multiplication. Previously, they might have looked at diagrams and patterns, written/recited tables and completed a number of consolidation questions.

Educational policies seem to change with the political wind and advice given about teaching numeracy can lead to some children thinking that *knowing* their tables simply involves reciting an answer pattern like 4, 8, 12, 16 and so on. My experience has lead me to notice that pupils who have been taught like that, initially, do less well on these worksheets and, in complete contrast, those pupils who know both the question and its answer scored better and in significantly faster times. Through regular use of the sheets over several months, I have witnessed this gap narrow to the point that most pupils could get full marks in a reasonable time. The challenge then becomes to do them faster and faster.

This book contains two sets of structured photocopiable sheets that can be used over and over again, not just within the same year but also year after year. The repetition enables a teacher to encourage their pupils to steadily improve, especially on their previous Climb The Wall results, by recognising and praising what they have done well and identifying what still needs to be learnt. This review process allows the setting of simple learning targets either by the teacher or the pupils themselves. By practising their areas of weakness, this should help them to get a higher total or a faster time on the next occasion.

The Build The Wall sheets test straight forward recall of number facts for each individual table from 2 to 10. The only exception is Build the Wall 1 which tests the skill of multiplying by both 0 and 1 on the same sheets. I have created 2 worksheet versions (a and b) for each table from 2 to 9 to allow for repetition when checking or for consolidation. These worksheets are a simple resource to help younger or less able pupils to practise their knowledge of times tables.

Over time, most pupils should be able to do these well and so might *appear* to know their tables. In order to check this, the Climb The Wall sheets test knowledge of 3 tables at a time and there are 3 versions (a, b and c) of each. To complete the set of worksheets, Climb The Wall 9 tests recall of the 6, 7, 8 and 9 times tables and there are 5 different versions available to use.

When creating these sheets, it was not my intention for all pupils to start on Climb The Wall 1a and monotonously work through each sheet until completing Climb The Wall 9e. The teacher, or support staff, should identify which tables the pupil(s) need to improve upon and then select the appropriate sheets for that task.

When pupils are proficient at Climb The Wall sheets 9a to 9e then they are ready to move onto The Wall sheets which test recall of all 10 tables at the same time. These can be found in my Beat The Wall book that was written mainly for upper primary, middle and secondary/high school pupils.

The worksheets in my books have been presented in letter size (approx A4) as this makes the numbers clear for pupils to read. However, in order to save photocopying expense it might be possible to reduce each sheet down to A5 and then produce working masters that have 2 sheets to a page so long as your pupils are still able to read the numbers clearly at that size.

I mostly used these resources as a starter as they quickly got the class to settle and focus on practising their numeracy skills. The sheets could also be used as a filler or break between 2 tasks, especially in longer maths lessons. Sometimes finding work for cover lessons can be a problem but once the pupils and staff know the routine associated with these sheets then they can be an appropriate and useful task for around 15 to 20 minutes of that lesson. Setting them as a piece of homework is also possible so long as the children can be trusted not to use a calculator to answer all of the questions.

Initially you might need to allow 5 to 15 minutes of a lesson to complete a sheet, reducing this as the children get better. If you choose to time the pupils, how you achieve this is up to you and your classroom management. Where the whole class are doing a sheet at the same time, a digital stopwatch could be projected onto a smartboard for the pupils to look at and note when they finish. Where appropriate, a stopclock could be put on each worktable for small groups to share. As I often taught lessons in different rooms, it was simpler for me to use a timer on a digital watch or a smartphone and call out times when pupils said, "Done" or "Finished".

When checking answers, several possibilities could be employed. Choosing which one works best depends on how each individual teacher organises their class and classroom resources.

- Copies of the answer sheets could be made available and pupils could self-mark
- Pupils could swap papers and mark their partner's work from the available answer sheets
- Support staff could work with small groups of pupils, monitoring and discussing issues as they arise when going through the answers
- The teacher or support staff could mark all of the sheets and set targets before returning them
- The teacher could read out the answers for a specific worksheet to the whole class at the same time (with or without swapping papers).

An advantage of the last method is that it allows the teacher to interact with the pupils about certain questions as they move through the sheet. Perhaps asking questions like;

- How do we know that 63 divides by 9? (divisibility rule 6 + 3 = 9)

- How can we tell if a large number divides by 5? Or by 3? Or by 6?
- Who can show me a way to do that question?
- Can anyone do it a different way?
- What type of number is 25?
- What is the square root of 49?
- What do we mean by multiple? Or factor?
- Who can give me an estimate of the answer?
- Who can give me a real-life problem where that question is what we need to work out?
- Do you think the answer is going to be bigger or smaller than…?
- If I think the answer is 24. Who can tell me why that can't be right?

Other memory tricks used to help with tables in my last school were;

8 x 8 = 64	I ate and I ate and I was sick on the floor
	$\quad$ 8 $\qquad$ 8 $\qquad$ 6 $\qquad$ 4
56 = 7 x 8	five, six, seven, eight
6 x 7 = 42	I was sick in Devon for 2 days
	$\quad$ 6 $\qquad$ 7 $\quad$ 4 2

Saying these as the appropriate answers came up often helped those who struggled particularly with their 6, 7 and 8 times tables. I'm sure there are other tricks that you or your pupils can devise. Using the answer session to widen the focus into other areas of the curriculum can help the pupils to see the links that occur in maths rather than seeing each topic area in isolation.

Once a sheet has been marked, it can be used to identify errors. Staff, or pupils themselves, could circle 3, 4 or perhaps 5 mistakes as targets. Pupils could then take the sheet home and learn the correct answers for next time. This might also be a way to involve parents in the process.

Where a pupil has not yet achieved full marks for a sheet then an individual target of getting *at least* 1 or 2 additional marks next time might be sufficient, especially with those who struggle to make progress. If the target is set too challenging, e.g. more than 5, then the pupil might make some progress but still feel disappointed because they have failed to reach their goal. By setting appropriate achievable targets, it is possible to continually praise pupils for their progress and this encourages them to keep working towards the satisfying achievement of "full marks".

Similar ideas apply to setting time targets, if you feel they are appropriate for your class, perhaps completing a sheet 5 or 10 seconds faster is often suitable. A time target is incredibly useful for those who regularly get full marks. The challenge now becomes to do it as quickly as they can and this then engages the most able pupils in the class as they try to get their best *personal* time or battle with others of a similar ability to hold the record for the fastest time.

What is a good time for the Climb The Wall sheets? Correctly completing one in 2 minutes or less shows a sound recall of their tables and it is even possible for the very best to achieve around 1 minute or under. A time of between 30 and 45 seconds would be impressive indeed!

Generally, a good time depends on which sheet is being done, the age and the ability of the pupils doing them. It is up to the professional judgement of the teacher to assess within the context of their class what constitutes a good or an amazing time.

When pupils have done particularly well with their score or time, this could be recognised by putting their name on the "Star of the Week" board or by giving them a reward of a star, merit, house point or other system that is used within your school.

In conclusion, simply handing out the sheets, giving the class a set time to do the questions and then reading out the answers is not in itself going to make much headway in learning their tables. The sheets need to be set regularly, scores and times need to be recorded, preferably in the pupils' workbooks or folders so they have easy access to what they did last time and the targets they are trying to achieve next time. The teacher should spend some time after each sheet collecting scores so they can discuss progress made and interact with the pupils about their new targets. This shows that the teacher values the process and it encourages the pupils to do better next time.

I cannot stress highly enough how important the use of targets, praise and encouragement is in helping pupils who struggle to improve their own weak areas in numeracy. Build The Wall and Climb The Wall worksheets, if used regularly and as explained, will help a wide range of children to achieve better scores and better times. Using these resources will not only inform teachers about pupil achievement but also they will help struggling pupils to recognise that they have made progress over time and ultimately that should make them feel positive about themselves, about doing maths and about beating that *wall*.

Tony Colledge

Build The Wall

Build The Wall 1 tests the recall of multiplying by 0 and 1

Build The Wall 2 to Build The Wall 10 test the recall of multiplication facts for each times table from 2 to 10

Answers are provided at the end of this section.

An example of a pupil record sheet is provided with the answers.

NAME: _____

BUILD THE WALL

1a

KEY FOCUS: *Multiplying by 0 and 1*

How many did you get right ?

How fast did you do it ?

SCORE	/ 22
mins	**secs**

YOU CAN DO IT !

$0 \times 6 =$

$0 \times 1 =$ $1 \times 4 =$

$1 \times 7 =$ $0 \times 5 =$

$0 \times 10 =$ $2 \times 1 =$ $3 \times 0 =$

$1 \times 0 =$ $6 \times 1 =$ $0 \times 0 =$

$9 \times 1 =$ $7 \times 0 =$ $1 \times 1 =$ $0 \times 8 =$

$1 \times 5 =$ $0 \times 9 =$ $1 \times 10 =$

$0 \times 2 =$ $3 \times 1 =$ $1 \times 8 =$ $4 \times 0 =$

NAME: _____

BUILD THE WALL

1b

KEY FOCUS: *Multiplying by 0 and 1*

How many did you get right ?

How fast did you do it ?

SCORE	/ 22
mins	secs

YOU CAN DO IT !

$9 \times 0 =$

$1 \times 2 =$ $1 \times 0 =$

$1 \times 6 =$ $0 \times 10 =$

$0 \times 1 =$ $0 \times 0 =$ $1 \times 4 =$

$3 \times 1 =$ $8 \times 0 =$ $1 \times 10 =$

$2 \times 0 =$ $7 \times 1 =$ $0 \times 5 =$ $1 \times 8 =$

$6 \times 0 =$ $1 \times 1 =$ $0 \times 4 =$

$5 \times 1 =$ $3 \times 0 =$ $0 \times 7 =$ $1 \times 9 =$

BUILD THE WALL

2a

<u>KEY FOCUS</u>: *Multiplying by 2*

How many did you get right ?

How fast did you do it ?

NAME: _____

SCORE	22
mins	secs

YOU CAN DO IT !

8 x 2 =

3 x 2 =   2 x 5 =

2 x 9 =   0 x 2 =

2 x 2 =   10 x 2 =   2 x 4 =

5 x 2 =   2 x 7 =   1 x 2 =

2 x 8 =   2 x 0 =   2 x 3 =   6 x 2 =

9 x 2 =   7 x 2 =   2 x 2 =

2 x 10 =   2 x 6 =   2 x 1 =   4 x 2 =

NAME: _____

BUILD THE WALL

2b

KEY FOCUS: *Multiplying by 2*

How many did you get right ?

How fast did you do it ?

SCORE	22
mins	secs

YOU CAN DO IT !

2 x 7 =

2 x 9 =   2 x 2 =

10 x 2 =   2 x 3 =

2 x 6 =   0 x 2 =   2 x 1 =

4 x 2 =   2 x 8 =   2 x 5 =

5 x 2 =   1 x 2 =   3 x 2 =   7 x 2 =

2 x 10 =   2 x 4 =   8 x 2 =

6 x 2 =   2 x 2 =   2 x 0 =   9 x 2 =

NAME: _____

BUILD THE WALL

KEY FOCUS: *Multiplying by 3*

3a

How many did you get right ?

How fast did you do it ?

SCORE	22
mins	secs

YOU CAN DO IT !

6 x 3 =

10 x 3 =   3 x 3 =

2 x 3 =   3 x 8 =

3 x 4 =   9 x 3 =   1 x 3 =

7 x 3 =   0 x 3 =   3 x 5 =

3 x 10 =   3 x 2 =   3 x 6 =   4 x 3 =

5 x 3 =   3 x 7 =   3 x 1 =

3 x 9 =   3 x 0 =   8 x 3 =   3 x 3 =

NAME: _____

BUILD THE WALL

3b

KEY FOCUS: *Multiplying by 3*

How many did you get right ?

How fast did you do it ?

SCORE	22
mins	secs

YOU CAN DO IT !

3 x 3 =

5 x 3 =   3 x 10 =

7 x 3 =   3 x 2 =

9 x 3 =   3 x 1 =   4 x 3 =

0 x 3 =   3 x 8 =   3 x 6 =

2 x 3 =   3 x 7 =   3 x 3 =   3 x 9 =

3 x 4 =   8 x 3 =   3 x 0 =

10 x 3 =   3 x 5 =   1 x 3 =   6 x 3 =

BUILD THE WALL

KEY FOCUS: *Multiplying by 4*

4a

How many did you get right ?

How fast did you do it ?

SCORE	/ 22
mins	secs

YOU CAN DO IT !

4 x 5 =

4 x 8 =   3 x 4 =

4 x 0 =   4 x 6 =

4 x 4 =   4 x 9 =   4 x 2 =

4 x 10 =   1 x 4 =   7 x 4 =

5 x 4 =   8 x 4 =   0 x 4 =   4 x 4 =

4 x 7 =   2 x 4 =   10 x 4 =

6 x 4 =   4 x 3 =   9 x 4 =   4 x 1 =

BUILD THE WALL

4b

KEY FOCUS: *Multiplying by 4*

How many did you get right ?

How fast did you do it ?

SCORE / 22

mins   secs

YOU CAN DO IT !

$9 \times 4 =$

$0 \times 4 =$ $4 \times 8 =$

$10 \times 4 =$ $4 \times 4 =$

$4 \times 3 =$ $7 \times 4 =$ $4 \times 5 =$

$8 \times 4 =$ $2 \times 4 =$ $4 \times 1 =$

$6 \times 4 =$ $1 \times 4 =$ $4 \times 10 =$ $3 \times 4 =$

$4 \times 2 =$ $4 \times 7 =$ $4 \times 0 =$

$4 \times 4 =$ $4 \times 9 =$ $5 \times 4 =$ $4 \times 6 =$

NAME: _____

BUILD THE WALL

5a

KEY FOCUS: *Multiplying by 5*

How many did you get right ?

How fast did you do it ?

SCORE	22
mins	secs

YOU CAN DO IT !

5 x 3 =

5 x 5 =   10 x 5 =

5 x 8 =   2 x 5 =

5 x 6 =   5 x 1 =   7 x 5 =

0 x 5 =   9 x 5 =   5 x 4 =

5 x 7 =   3 x 5 =   5 x 10 =   5 x 5 =

4 x 5 =   5 x 9 =   1 x 5 =

6 x 5 =   5 x 2 =   5 x 0 =   8 x 5 =

NAME: _____

BUILD THE WALL

5b

KEY FOCUS: *Multiplying by 5*

How many did you get right ?

How fast did you do it ?

SCORE	/ 22
mins	secs

YOU CAN DO IT !

7 x 5 =

10 x 5 =   0 x 5 =

5 x 2 =   5 x 8 =

5 x 9 =   5 x 5 =   3 x 5 =

4 x 5 =   1 x 5 =   5 x 6 =

5 x 3 =   6 x 5 =   5 x 0 =   5 x 10 =

5 x 5 =   8 x 5 =   5 x 4 =

9 x 5 =   2 x 5 =   5 x 7 =   5 x 1 =

NAME: _____

BUILD THE WALL

6a

KEY FOCUS: *Multiplying by 6*

How many did you get right ?

How fast did you do it ?

SCORE	/ 22
mins	secs

YOU CAN DO IT !

5 x 6 =

7 x 6 =   6 x 2 =

0 x 6 =   9 x 6 =

6 x 3 =   6 x 10 =   6 x 1 =

6 x 6 =   4 x 6 =   6 x 8 =

6 x 7 =   10 x 6 =   6 x 0 =   6 x 5 =

6 x 4 =   6 x 9 =   1 x 6 =

2 x 6 =   6 x 6 =   8 x 6 =   3 x 6 =

NAME: _____

BUILD THE WALL

6b

KEY FOCUS: *Multiplying by 6*

How many did you get right ?

How fast did you do it ?

SCORE	/ 22
mins	**secs**

YOU CAN DO IT !

$9 \times 6 =$

$2 \times 6 =$

$6 \times 5 =$

$10 \times 6 =$

$6 \times 3 =$

$6 \times 1 =$

$6 \times 8 =$

$4 \times 6 =$

$5 \times 6 =$

$7 \times 6 =$

$6 \times 2 =$

$6 \times 6 =$

$0 \times 6 =$

$6 \times 10 =$

$1 \times 6 =$

$8 \times 6 =$

$6 \times 4 =$

$6 \times 7 =$

$3 \times 6 =$

$6 \times 9 =$

$6 \times 6 =$

$6 \times 0 =$

NAME: _____

BUILD THE WALL

7a

KEY FOCUS: *Multiplying by 7*

How many did you get right ?

How fast did you do it ?

SCORE	22
mins	secs

YOU CAN DO IT !

6 x 7 =

2 x 7 =   7 x 9 =

7 x 0 =   7 x 7 =

7 x 8 =   3 x 7 =   7 x 10 =

4 x 7 =   7 x 7 =   1 x 7 =

7 x 5 =   0 x 7 =   9 x 7 =   7 x 3 =

10 x 7 =   7 x 2 =   7 x 6 =

7 x 4 =   7 x 1 =   8 x 7 =   5 x 7 =

NAME: _____

BUILD THE WALL

7b

KEY FOCUS: *Multiplying by 7*

How many did you get right ?

How fast did you do it ?

SCORE	/ 22
mins	secs

YOU CAN DO IT !

9 x 7 =

10 x 7 =   4 x 7 =

7 x 1 =   7 x 6 =

5 x 7 =   7 x 8 =   7 x 2 =

7 x 3 =   7 x 7 =   7 x 0 =

1 x 7 =   7 x 10 =   2 x 7 =   7 x 5 =

8 x 7 =   7 x 4 =   6 x 7 =

7 x 7 =   0 x 7 =   7 x 9 =   3 x 7 =

NAME: _____

BUILD THE WALL

KEY FOCUS: *Multiplying by 8*

8a

How many did you get right ?

How fast did you do it ?

SCORE	22
mins	secs

YOU CAN DO IT !

$8 \times 4 =$

$8 \times 9 =$ $5 \times 8 =$

$2 \times 8 =$ $8 \times 7 =$

$8 \times 3 =$ $10 \times 8 =$ $8 \times 1 =$

$8 \times 6 =$ $0 \times 8 =$ $8 \times 8 =$

$7 \times 8 =$ $4 \times 8 =$ $8 \times 2 =$ $9 \times 8 =$

$1 \times 8 =$ $8 \times 10 =$ $8 \times 5 =$

$3 \times 8 =$ $8 \times 8 =$ $8 \times 0 =$ $6 \times 8 =$

NAME: _____

BUILD THE WALL

8b

KEY FOCUS: *Multiplying by 8*

How many did you get right ?

How fast did you do it ?

SCORE	/ 22
mins	secs

YOU CAN DO IT !

9 x 8 =

8 x 6 =   8 x 0 =

10 x 8 =   8 x 3 =

2 x 8 =   8 x 8 =   5 x 8 =

1 x 8 =   7 x 8 =   8 x 4 =

8 x 5 =   0 x 8 =   8 x 10 =   3 x 8 =

8 x 8 =   8 x 2 =   6 x 8 =

4 x 8 =   8 x 7 =   8 x 1 =   8 x 9 =

NAME: _____

BUILD THE WALL

9a

KEY FOCUS: *Multiplying by 9*

How many did you get right ?

How fast did you do it ?

SCORE	/ 22
mins	secs

YOU CAN DO IT !

9 x 7 =

4 x 9 =   10 x 9 =

9 x 8 =   1 x 9 =

3 x 9 =   9 x 5 =   9 x 9 =

7 x 9 =   9 x 2 =   9 x 0 =

9 x 9 =   9 x 1 =   5 x 9 =   9 x 6 =

8 x 9 =   9 x 3 =   0 x 9 =

9 x 4 =   6 x 9 =   2 x 9 =   9 x 10 =

NAME: _____

BUILD THE WALL

9b

© 2014 Tony Colledge   <u>KEY FOCUS</u>: *Multiplying by 9*

How many did you get right ?

How fast did you do it ?

SCORE	/ 22
mins	secs

YOU CAN DO IT !

9 x 9 =

7 x 9 =   9 x 5 =

9 x 4 =   10 x 9 =

3 x 9 =   9 x 8 =   0 x 9 =

9 x 6 =   1 x 9 =   9 x 9 =

4 x 9 =   9 x 10 =   9 x 2 =   5 x 9 =

9 x 1 =   6 x 9 =   9 x 3 =

8 x 9 =   9 x 0 =   9 x 7 =   2 x 9 =

NAME: _____

BUILD THE WALL

KEY FOCUS: *Multiplying by 10*

10

SCORE	22
mins	secs

How many did you get right ?

How fast did you do it ?

YOU CAN DO IT !

$10 \times 0 =$

$8 \times 10 =$ $10 \times 6 =$

$10 \times 2 =$ $10 \times 10 =$

$5 \times 10 =$ $10 \times 7 =$ $1 \times 10 =$

$6 \times 10 =$ $4 \times 10 =$ $10 \times 9 =$

$3 \times 10 =$ $10 \times 10 =$ $0 \times 10 =$ $10 \times 5 =$

$10 \times 8 =$ $2 \times 10 =$ $10 \times 4 =$

$10 \times 3 =$ $9 \times 10 =$ $10 \times 1 =$ $7 \times 10 =$

Build The Wall

ANSWERS

DATE	TASK	SCORE	TIME	TARGET(S)

ANSWERS

BUILD THE WALL 1a

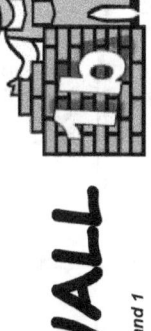

KEY FOCUS: *Multiplying by 0 and 1*

What score/time did you get? Are you pleased with it?
What could you do to improve your score or time?
What work do you need to do now?

WELL DONE. KEEP IT UP!

0 x 6 = 0 | 1 x 4 = 4
0 x 1 = 0 | 0 x 5 = 0 | 3 x 0 = 0
1 x 7 = 7 | 2 x 1 = 2 | 0 x 0 = 0 | 0 x 8 = 0
0 x 10 = 0 | 6 x 1 = 6 | 1 x 1 = 1 | 1 x 10 = 10 | 4 x 0 = 0
1 x 0 = 0 | 7 x 0 = 0 | 0 x 9 = 0 | 1 x 8 = 8
9 x 1 = 9 | 1 x 5 = 5 | 3 x 1 = 3
0 x 2 = 0

ANSWERS

BUILD THE WALL 1b

KEY FOCUS: *Multiplying by 0 and 1*

What score/time did you get? Are you pleased with it?
What could you do to improve your score or time?
What work do you need to do now?

WELL DONE. KEEP IT UP!

9 x 0 = 0
1 x 2 = 2 | 1 x 0 = 0
1 x 6 = 6 | 0 x 10 = 0
0 x 1 = 0 | 0 x 0 = 0 | 1 x 4 = 4
3 x 1 = 3 | 8 x 0 = 0 | 1 x 10 = 10 | 1 x 8 = 8
2 x 0 = 0 | 7 x 1 = 7 | 0 x 5 = 0 | 0 x 4 = 0
6 x 0 = 0 | 1 x 1 = 1
5 x 1 = 5 | 3 x 0 = 0 | 0 x 7 = 0 | 1 x 9 = 9

ANSWERS

BUILD THE WALL 2b

KEY FOCUS: *Multiplying by 2*

What score/time did you get ?   Are you pleased with it ?
What could you do to improve your score or time ?
What work do you need to do now ?

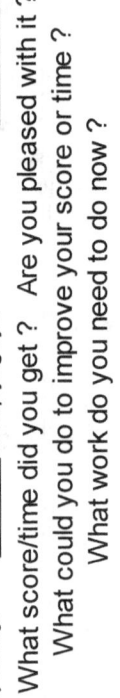

WELL DONE.
KEEP IT UP !

2 x 7 = **14**   2 x 2 = **4**

2 x 9 = **18**   2 x 3 = **6**

10 x 2 = **20**   0 x 2 = **0**   2 x 1 = **2**

2 x 6 = **12**   2 x 5 = **10**

4 x 2 = **8**   2 x 8 = **16**   7 x 2 = **14**

5 x 2 = **10**   1 x 2 = **2**   3 x 2 = **6**   8 x 2 = **16**

2 x 10 = **20**   2 x 4 = **8**

6 x 2 = **12**   2 x 2 = **4**   2 x 0 = **0**   9 x 2 = **18**

ANSWERS

BUILD THE WALL 2a

KEY FOCUS: *Multiplying by 2*

What score/time did you get ?   Are you pleased with it ?
What could you do to improve your score or time ?
What work do you need to do now ?

WELL DONE.
KEEP IT UP !

8 x 2 = **16**   2 x 5 = **10**

3 x 2 = **6**   0 x 2 = **0**   2 x 4 = **8**

2 x 9 = **18**   10 x 2 = **20**   1 x 2 = **2**   6 x 2 = **12**

2 x 7 = **14**   2 x 3 = **6**   2 x 2 = **4**

2 x 0 = **0**   7 x 2 = **14**   2 x 1 = **2**   4 x 2 = **8**

5 x 2 = **10**   2 x 6 = **12**

9 x 2 = **18**

2 x 8 = **16**

2 x 10 = **20**

ANSWERS

BUILD THE WALL 3b

KEY FOCUS: *Multiplying by 3*

What score/time did you get? Are you pleased with it?
What could you do to improve your score or time?
What work do you need to do now?

WELL DONE.
KEEP IT UP !

Wall answers:

$3 \times 3 = 9$, $3 \times 10 = 30$, $3 \times 2 = 6$, $4 \times 3 = 12$
$5 \times 3 = 15$, $3 \times 6 = 18$, $3 \times 9 = 27$, $6 \times 3 = 18$
$7 \times 3 = 21$, $3 \times 1 = 3$, $3 \times 3 = 9$, $3 \times 0 = 0$
$9 \times 3 = 27$, $3 \times 8 = 24$, $8 \times 3 = 24$, $1 \times 3 = 3$
$0 \times 3 = 0$, $3 \times 7 = 21$, $3 \times 5 = 15$
$2 \times 3 = 6$, $3 \times 4 = 12$, $10 \times 3 = 30$

ANSWERS

BUILD THE WALL 3a

KEY FOCUS: *Multiplying by 3*

What score/time did you get? Are you pleased with it?
What could you do to improve your score or time?
What work do you need to do now?

WELL DONE.
KEEP IT UP !

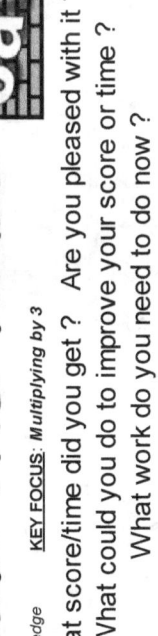

Wall answers:

$6 \times 3 = 18$, $3 \times 3 = 9$
$10 \times 3 = 30$, $3 \times 8 = 24$, $1 \times 3 = 3$
$2 \times 3 = 6$, $9 \times 3 = 27$, $3 \times 5 = 15$, $4 \times 3 = 12$
$3 \times 4 = 12$, $0 \times 3 = 0$, $3 \times 6 = 18$, $3 \times 1 = 3$
$7 \times 3 = 21$, $3 \times 2 = 6$, $3 \times 7 = 21$, $8 \times 3 = 24$, $3 \times 3 = 9$
$3 \times 10 = 30$, $5 \times 3 = 15$, $3 \times 0 = 0$
$3 \times 9 = 27$

ANSWERS

BUILD THE WALL 4b

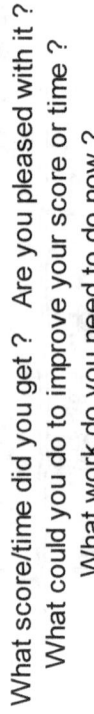

KEY FOCUS: *Multiplying by 4*

What score/time did you get ?   Are you pleased with it ?
What could you do to improve your score or time ?
What work do you need to do now ?

WELL DONE.
KEEP IT UP !

$9 \times 4 = 36$

$0 \times 4 = 0$ $4 \times 8 = 32$

$10 \times 4 = 40$ $4 \times 4 = 16$

$4 \times 3 = 12$ $7 \times 4 = 28$ $4 \times 5 = 20$

$8 \times 4 = 32$ $2 \times 4 = 8$ $4 \times 1 = 4$

$1 \times 4 = 4$ $4 \times 10 = 40$ $3 \times 4 = 12$

$6 \times 4 = 24$ $4 \times 2 = 8$ $4 \times 7 = 28$ $4 \times 0 = 0$

$4 \times 4 = 16$ $4 \times 9 = 36$ $5 \times 4 = 20$ $4 \times 6 = 24$

ANSWERS

BUILD THE WALL 4a

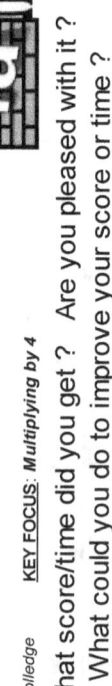

KEY FOCUS: *Multiplying by 4*

What score/time did you get ?   Are you pleased with it ?
What could you do to improve your score or time ?
What work do you need to do now ?

WELL DONE.
KEEP IT UP !

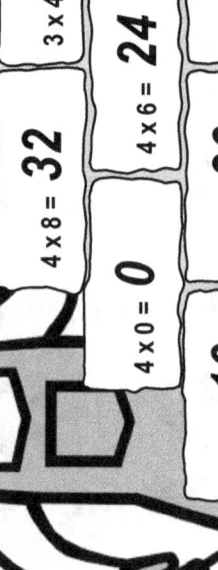

$4 \times 5 = 20$

$4 \times 8 = 32$ $3 \times 4 = 12$

$4 \times 0 = 0$ $4 \times 6 = 24$

$4 \times 4 = 16$ $4 \times 9 = 36$ $4 \times 2 = 8$

$1 \times 4 = 4$ $7 \times 4 = 28$

$8 \times 4 = 32$ $0 \times 4 = 0$ $4 \times 4 = 16$

$4 \times 10 = 40$ $4 \times 7 = 28$ $2 \times 4 = 8$ $10 \times 4 = 40$

$5 \times 4 = 20$ $4 \times 3 = 12$ $9 \times 4 = 36$ $4 \times 1 = 4$

$6 \times 4 = 24$

ANSWERS

BUILD THE WALL 5b

KEY FOCUS: *Multiplying by 5*

What score/time did you get ?   Are you pleased with it ?
What could you do to improve your score or time ?
What work do you need to do now ?

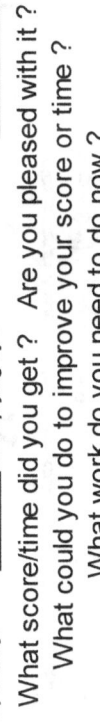

WELL DONE.
KEEP IT UP !

$7 \times 5 = 35$

$10 \times 5 = 50$ $5 \times 2 = 10$ $5 \times 9 = 45$ $4 \times 5 = 20$

$0 \times 5 = 0$ $5 \times 8 = 40$ $3 \times 5 = 15$

$5 \times 5 = 25$ $1 \times 5 = 5$ $6 \times 5 = 30$ $5 \times 3 = 15$

$5 \times 6 = 30$ $5 \times 0 = 0$ $5 \times 10 = 50$

$5 \times 4 = 20$ $8 \times 5 = 40$ $5 \times 5 = 25$ $9 \times 5 = 45$

$5 \times 1 = 5$ $5 \times 7 = 35$ $2 \times 5 = 10$

ANSWERS

BUILD THE WALL 5a

KEY FOCUS: *Multiplying by 5*

What score/time did you get ?   Are you pleased with it ?
What could you do to improve your score or time ?
What work do you need to do now ?

WELL DONE.
KEEP IT UP !

$5 \times 3 = 15$

$5 \times 5 = 25$ $5 \times 8 = 40$ $5 \times 6 = 30$ $0 \times 5 = 0$

$10 \times 5 = 50$ $2 \times 5 = 10$ $5 \times 4 = 20$ $5 \times 5 = 25$ $8 \times 5 = 40$

$5 \times 1 = 5$ $9 \times 5 = 45$ $3 \times 5 = 15$ $5 \times 7 = 35$

$7 \times 5 = 35$ $5 \times 10 = 50$ $5 \times 9 = 45$ $4 \times 5 = 20$

$1 \times 5 = 5$ $5 \times 0 = 0$ $5 \times 2 = 10$ $6 \times 5 = 30$

ANSWERS

BUILD THE WALL 6b

KEY FOCUS: *Multiplying by 6*

What score/time did you get ? Are you pleased with it ?
What could you do to improve your score or time ?
What work do you need to do now ?

WELL DONE.
KEEP IT UP !

9 x 6 = 54

6 x 5 = 30
2 x 6 = 12
10 x 6 = 60
6 x 1 = 6
5 x 6 = 30

6 x 3 = 18
6 x 8 = 48
7 x 6 = 42
0 x 6 = 0
6 x 6 = 36

4 x 6 = 24
6 x 2 = 12
6 x 10 = 60
6 x 4 = 24
8 x 6 = 48

1 x 6 = 6
6 x 7 = 42
6 x 6 = 36
6 x 9 = 54
3 x 6 = 18

6 x 0 = 0

ANSWERS

BUILD THE WALL 6a

KEY FOCUS: *Multiplying by 6*

What score/time did you get ? Are you pleased with it ?
What could you do to improve your score or time ?
What work do you need to do now ?

WELL DONE.
KEEP IT UP !

5 x 6 = 30
6 x 2 = 12
7 x 6 = 42
9 x 6 = 54
0 x 6 = 0
6 x 10 = 60
6 x 0 = 0
6 x 1 = 6
6 x 8 = 48
1 x 6 = 6
6 x 5 = 30

6 x 3 = 18
4 x 6 = 24
10 x 6 = 60
6 x 9 = 54
8 x 6 = 48
3 x 6 = 18

6 x 6 = 36
6 x 4 = 24
6 x 6 = 36

6 x 7 = 42
2 x 6 = 12

ANSWERS

BUILD THE WALL 7a

KEY FOCUS: *Multiplying by 7*

What score/time did you get ?   Are you pleased with it ?
What could you do to improve your score or time ?
What work do you need to do now ?

WELL DONE.
KEEP IT UP !

6 x 7 = **42**
7 x 9 = **63**

2 x 7 = **14**
7 x 7 = **49**
7 x 10 = **70**

7 x 0 = **0**
3 x 7 = **21**
1 x 7 = **7**
7 x 3 = **21**

7 x 8 = **56**
7 x 7 = **49**
9 x 7 = **63**
7 x 6 = **42**
5 x 7 = **35**

4 x 7 = **28**
0 x 7 = **0**
7 x 2 = **14**
8 x 7 = **56**

7 x 5 = **35**
10 x 7 = **70**
7 x 1 = **7**

7 x 4 = **28**

ANSWERS

BUILD THE WALL 7b

KEY FOCUS: *Multiplying by 7*

What score/time did you get ?   Are you pleased with it ?
What could you do to improve your score or time ?
What work do you need to do now ?

WELL DONE.
KEEP IT UP !

9 x 7 = **63**
4 x 7 = **28**

10 x 7 = **70**
7 x 6 = **42**
7 x 2 = **14**

7 x 1 = **7**
7 x 8 = **56**
7 x 0 = **0**
7 x 5 = **35**

5 x 7 = **35**
7 x 7 = **49**
2 x 7 = **14**
6 x 7 = **42**
3 x 7 = **21**

7 x 3 = **21**
7 x 10 = **70**
7 x 4 = **28**
7 x 9 = **63**

1 x 7 = **7**
8 x 7 = **56**
0 x 7 = **0**

7 x 7 = **49**

ANSWERS

BUILD THE WALL 8b

KEY FOCUS: Multiplying by 8

What score/time did you get? Are you pleased with it?
What could you do to improve your score or time?
What work do you need to do now?

WELL DONE. KEEP IT UP !

$9 \times 8 = 72$

$8 \times 6 = 48$ | $8 \times 0 = 0$

$10 \times 8 = 80$ | $8 \times 3 = 24$

$2 \times 8 = 16$ | $8 \times 8 = 64$ | $5 \times 8 = 40$

$7 \times 8 = 56$ | $8 \times 4 = 32$

$0 \times 8 = 0$ | $8 \times 10 = 80$ | $3 \times 8 = 24$

$1 \times 8 = 8$ | $8 \times 2 = 16$ | $6 \times 8 = 48$

$8 \times 5 = 40$ | $8 \times 8 = 64$ | $8 \times 1 = 8$

$4 \times 8 = 32$ | $8 \times 7 = 56$ | $8 \times 9 = 72$

ANSWERS

BUILD THE WALL 8a

KEY FOCUS: Multiplying by 8

What score/time did you get? Are you pleased with it?
What could you do to improve your score or time?
What work do you need to do now?

WELL DONE. KEEP IT UP !

$8 \times 4 = 32$ | $5 \times 8 = 40$

$8 \times 9 = 72$ | $8 \times 7 = 56$ | $8 \times 1 = 8$

$2 \times 8 = 16$ | $10 \times 8 = 80$ | $8 \times 8 = 64$ | $9 \times 8 = 72$

$8 \times 3 = 24$ | $0 \times 8 = 0$ | $8 \times 2 = 16$ | $8 \times 5 = 40$

$8 \times 6 = 48$ | $4 \times 8 = 32$ | $8 \times 10 = 80$ | $8 \times 0 = 0$ | $6 \times 8 = 48$

$7 \times 8 = 56$ | $1 \times 8 = 8$ | $8 \times 8 = 64$

$3 \times 8 = 24$

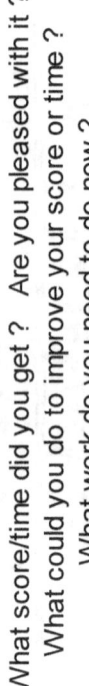

ANSWERS

BUILD THE WALL 9b

© 2014 Tony Colledge   KEY FOCUS: *Multiplying by 9*

What score/time did you get ?   Are you pleased with it ?
What could you do to improve your score or time ?
What work do you need to do now ?

WELL DONE.
KEEP IT UP !

$9 \times 9 = 81$
$9 \times 5 = 45$
$7 \times 9 = 63$
$10 \times 9 = 90$
$9 \times 4 = 36$
$9 \times 8 = 72$
$0 \times 9 = 0$
$3 \times 9 = 27$
$1 \times 9 = 9$
$9 \times 9 = 81$
$9 \times 6 = 54$
$9 \times 10 = 90$
$9 \times 2 = 18$
$5 \times 9 = 45$
$6 \times 9 = 54$
$9 \times 3 = 27$
$9 \times 1 = 9$
$9 \times 0 = 0$
$9 \times 7 = 63$
$4 \times 9 = 36$
$8 \times 9 = 72$
$2 \times 9 = 18$

ANSWERS

BUILD THE WALL 9a

© 2014 Tony Colledge   KEY FOCUS: *Multiplying by 9*

What score/time did you get ?   Are you pleased with it ?
What could you do to improve your score or time ?
What work do you need to do now ?

WELL DONE.
KEEP IT UP !

$9 \times 7 = 63$
$10 \times 9 = 90$
$4 \times 9 = 36$
$1 \times 9 = 9$
$9 \times 9 = 81$
$9 \times 8 = 72$
$9 \times 5 = 45$
$9 \times 2 = 18$
$5 \times 9 = 45$
$9 \times 0 = 0$
$9 \times 6 = 54$
$9 \times 3 = 27$
$0 \times 9 = 0$
$2 \times 9 = 18$
$9 \times 10 = 90$
$3 \times 9 = 27$
$9 \times 1 = 9$
$7 \times 9 = 63$
$9 \times 2 = 18$
$8 \times 9 = 72$
$6 \times 9 = 54$
$9 \times 9 = 81$
$9 \times 4 = 36$

ANSWERS

BUILD THE WALL

KEY FOCUS: *Multiplying by 10*

10

What score/time did you get ? Are you pleased with it ?
What could you do to improve your score or time ?
What work do you need to do now ?

WELL DONE.

KEEP IT UP !

10 x 0 = **0**		
8 x 10 = **80**	10 x 6 = **60**	
10 x 2 = **20**	10 x 10 = **100**	
5 x 10 = **50**	10 x 7 = **70**	1 x 10 = **10**
4 x 10 = **40**	10 x 9 = **90**	
10 x 10 = **100**	0 x 10 = **0**	10 x 5 = **50**
2 x 10 = **20**	10 x 4 = **40**	
9 x 10 = **90**	10 x 1 = **10**	
10 x 3 = **30**	7 x 10 = **70**	
6 x 10 = **60**		
10 x 8 = **80**		
3 x 10 = **30**		

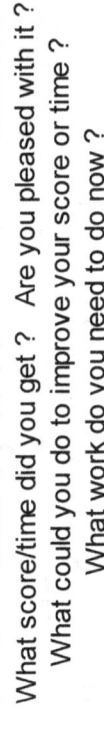

Climb The Wall

Climb The Wall 1 tests recall when multiplying by 2, 5 and 10
Climb The Wall 2 tests recall when multiplying by 2, 3 and 4
Climb The Wall 3 tests recall when multiplying by 3, 4 and 5
Climb The Wall 4 tests recall when multiplying by 4, 5 and 6
Climb The Wall 5 tests recall when multiplying by 3, 6 and 9
Climb The Wall 6 tests recall when multiplying by 4, 7 and 8
Climb The Wall 7 tests recall when multiplying by 6, 7 and 8
Climb The Wall 8 tests recall when multiplying by 7, 8 and 9
Climb The Wall 9 tests recall when multiplying by 6, 7, 8 and 9

Answers are provided at the end of this section.

An example of a pupil record sheet is provided with the answers.

NAME: _____

CLIMB THE WALL

KEY FOCUS: *Multiplying by 2, 5 and 10*

How many can you get right ?

How fast can you do it ?

What will you try to get next time ?

YOU CAN DO IT !

SCORE	/33
mins	secs
TARGETS	/33
mins	secs

$10 \times 10 =$

$10 \times 8 =$ $7 \times 10 =$

$5 \times 10 =$ $10 \times 0 =$ $3 \times 10 =$

$1 \times 10 =$ $10 \times 6 =$ $9 \times 10 =$ $10 \times 2 =$

$9 \times 5 =$ $5 \times 4 =$ $0 \times 5 =$ $2 \times 5 =$ $10 \times 4 =$

$5 \times 6 =$ $3 \times 5 =$ $5 \times 10 =$ $1 \times 5 =$

$2 \times 10 =$ $3 \times 2 =$ $5 \times 5 =$ $5 \times 8 =$ $7 \times 5 =$

$9 \times 2 =$ $2 \times 2 =$ $7 \times 2 =$ $2 \times 6 =$

$5 \times 2 =$ $1 \times 2 =$ $2 \times 8 =$ $0 \times 2 =$ $2 \times 4 =$

CLIMB THE WALL 1b

KEY FOCUS: *Multiplying by 2, 5 and 10*

How many can you get right ?

How fast can you do it ?

What will you try to get next time ?

YOU CAN DO IT !

SCORE	/33
mins	secs
TARGETS	/33
mins	secs

$8 \times 5 =$

$3 \times 10 =$ $5 \times 7 =$

$10 \times 4 =$ $2 \times 9 =$ $5 \times 0 =$

$2 \times 8 =$ $4 \times 2 =$ $10 \times 6 =$ $4 \times 5 =$

$5 \times 10 =$ $2 \times 0 =$ $5 \times 3 =$ $9 \times 5 =$ $1 \times 10 =$

$2 \times 10 =$ $6 \times 2 =$ $2 \times 5 =$ $10 \times 10 =$

$5 \times 5 =$ $2 \times 7 =$ $1 \times 2 =$ $7 \times 10 =$ $5 \times 2 =$

$5 \times 1 =$ $8 \times 10 =$ $2 \times 3 =$ $0 \times 10 =$

$10 \times 5 =$ $2 \times 2 =$ $9 \times 10 =$ $6 \times 5 =$ $10 \times 2 =$

NAME: _____

CLIMB THE WALL

KEY FOCUS: *Multiplying by 2, 5 and 10*

How many can you get right ?

How fast can you do it ?

What will you try to get next time ?

YOU CAN DO IT !

SCORE	/33
mins	secs
TARGETS	/33
mins	secs

$9 \times 5 =$

$5 \times 5 =$ $10 \times 10 =$

$0 \times 10 =$ $5 \times 10 =$ $4 \times 2 =$

$10 \times 2 =$ $5 \times 0 =$ $2 \times 1 =$ $4 \times 10 =$

$8 \times 10 =$ $2 \times 3 =$ $5 \times 6 =$ $2 \times 9 =$ $10 \times 1 =$

$5 \times 8 =$ $0 \times 2 =$ $10 \times 3 =$ $2 \times 7 =$

$1 \times 5 =$ $10 \times 7 =$ $6 \times 2 =$ $10 \times 9 =$ $5 \times 2 =$

$3 \times 5 =$ $2 \times 10 =$ $2 \times 2 =$ $7 \times 5 =$

$6 \times 10 =$ $2 \times 5 =$ $10 \times 5 =$ $8 \times 2 =$ $5 \times 4 =$

NAME: _____

CLIMB THE WALL 2a

KEY FOCUS: *Multiplying by 2, 3 and 4*

How many can you get right ?

How fast can you do it ?

What will you try to get next time ?

YOU CAN DO IT !

SCORE	/33
mins	secs
TARGETS	/33
mins	secs

$4 \times 4 =$

$10 \times 4 =$ $4 \times 3 =$

$0 \times 4 =$ $4 \times 5 =$ $8 \times 4 =$

$2 \times 4 =$ $4 \times 7 =$ $4 \times 1 =$ $4 \times 9 =$

$3 \times 3 =$ $9 \times 3 =$ $3 \times 4 =$ $3 \times 0 =$ $6 \times 4 =$

$3 \times 6 =$ $1 \times 3 =$ $3 \times 8 =$ $3 \times 2 =$

$2 \times 7 =$ $4 \times 2 =$ $5 \times 3 =$ $3 \times 10 =$ $7 \times 3 =$

$2 \times 1 =$ $2 \times 5 =$ $8 \times 2 =$ $2 \times 2 =$

$6 \times 2 =$ $0 \times 2 =$ $2 \times 9 =$ $2 \times 3 =$ $10 \times 2 =$

NAME: _____

CLIMB THE WALL 2b

KEY FOCUS: *Multiplying by 2, 3 and 4*

How many can you get right ?

How fast can you do it ?

What will you try to get next time ?

YOU CAN DO IT !

SCORE	/33
mins	secs
TARGETS	/33
mins	secs

$3 \times 7 =$

$4 \times 3 =$ $9 \times 4 =$

$5 \times 4 =$ $4 \times 10 =$ $3 \times 3 =$

$9 \times 2 =$ $3 \times 1 =$ $4 \times 8 =$ $2 \times 4 =$

$7 \times 4 =$ $2 \times 3 =$ $5 \times 2 =$ $6 \times 3 =$ $2 \times 0 =$

$2 \times 8 =$ $0 \times 3 =$ $2 \times 10 =$ $1 \times 4 =$

$4 \times 6 =$ $1 \times 2 =$ $3 \times 9 =$ $4 \times 4 =$ $3 \times 2 =$

$3 \times 4 =$ $10 \times 3 =$ $4 \times 2 =$ $2 \times 6 =$

$7 \times 2 =$ $2 \times 2 =$ $3 \times 5 =$ $4 \times 0 =$ $8 \times 3 =$

NAME: _____

CLIMB THE WALL 2c

KEY FOCUS: *Multiplying by 2, 3 and 4*

How many can you get right ?

How fast can you do it ?

What will you try to get next time ?

YOU CAN DO IT !

SCORE	/33
mins	secs
TARGETS	/33
mins	secs

$4 \times 9 =$

$5 \times 3 =$ $0 \times 2 =$

$2 \times 9 =$ $3 \times 3 =$ $2 \times 5 =$

$3 \times 0 =$ $6 \times 4 =$ $3 \times 10 =$ $4 \times 2 =$

$3 \times 6 =$ $2 \times 1 =$ $8 \times 2 =$ $3 \times 4 =$ $7 \times 3 =$

$10 \times 2 =$ $0 \times 4 =$ $6 \times 2 =$ $1 \times 3 =$

$4 \times 3 =$ $8 \times 4 =$ $4 \times 1 =$ $9 \times 3 =$ $3 \times 2 =$

$4 \times 7 =$ $2 \times 2 =$ $10 \times 4 =$ $2 \times 3 =$

$4 \times 5 =$ $3 \times 8 =$ $4 \times 4 =$ $2 \times 7 =$ $2 \times 4 =$

NAME: _____

CLIMB THE WALL

KEY FOCUS: *Multiplying by 3, 4 and 5*

How many can you get right ?

How fast can you do it ?

What will you try to get next time ?

YOU CAN DO IT !

SCORE	/33
mins	secs
TARGETS	/33
mins	secs

$10 \times 5 =$

$5 \times 5 =$ $2 \times 5 =$

$0 \times 5 =$ $8 \times 4 =$ $4 \times 5 =$

$6 \times 5 =$ $5 \times 3 =$ $5 \times 9 =$ $5 \times 1 =$

$4 \times 2 =$ $4 \times 10 =$ $4 \times 4 =$ $7 \times 4 =$ $5 \times 7 =$

$3 \times 4 =$ $4 \times 8 =$ $4 \times 0 =$ $4 \times 6 =$

$3 \times 5 =$ $10 \times 3 =$ $9 \times 4 =$ $1 \times 4 =$ $5 \times 4 =$

$6 \times 3 =$ $2 \times 3 =$ $3 \times 7 =$ $3 \times 3 =$

$0 \times 3 =$ $4 \times 3 =$ $3 \times 9 =$ $3 \times 1 =$ $8 \times 3 =$

NAME: _____

CLIMB THE WALL

3b

KEY FOCUS: *Multiplying by 3, 4 and 5*

How many can you get right ?

How fast can you do it ?

What will you try to get next time ?

YOU CAN DO IT !

SCORE	/33
mins	secs
TARGETS	/33
mins	secs

9 x 5 =

3 x 8 =   5 x 5 =

4 x 3 =   5 x 4 =   4 x 9 =

5 x 3 =   3 x 10 =   4 x 1 =   7 x 5 =

10 x 4 =   1 x 3 =   4 x 7 =   5 x 0 =   4 x 4 =

9 x 3 =   0 x 4 =   5 x 6 =   3 x 5 =

5 x 2 =   3 x 4 =   5 x 10 =   3 x 2 =   6 x 4 =

3 x 3 =   8 x 4 =   1 x 5 =   3 x 6 =

7 x 3 =   2 x 4 =   4 x 5 =   3 x 0 =   5 x 8 =

CLIMB THE WALL 3c

KEY FOCUS: *Multiplying by 3, 4 and 5*

How many can you get right ?

How fast can you do it ?

What will you try to get next time ?

YOU CAN DO IT !

SCORE	/33
mins	secs
TARGETS	/33
mins	secs

$7 \times 4 =$

$8 \times 5 =$ $4 \times 4 =$

$5 \times 3 =$ $9 \times 4 =$ $0 \times 5 =$

$3 \times 5 =$ $1 \times 4 =$ $10 \times 5 =$ $3 \times 4 =$

$4 \times 5 =$ $3 \times 9 =$ $2 \times 3 =$ $4 \times 8 =$ $5 \times 1 =$

$0 \times 3 =$ $4 \times 10 =$ $3 \times 3 =$ $6 \times 5 =$

$4 \times 2 =$ $8 \times 3 =$ $3 \times 1 =$ $5 \times 7 =$ $4 \times 3 =$

$6 \times 3 =$ $5 \times 4 =$ $10 \times 3 =$ $4 \times 0 =$

$5 \times 5 =$ $5 \times 9 =$ $3 \times 7 =$ $2 \times 5 =$ $4 \times 6 =$

NAME: _____

CLIMB THE WALL 4a

KEY FOCUS: *Multiplying by 4, 5 and 6*

How many can you get right ?

How fast can you do it ?

What will you try to get next time ?

YOU CAN DO IT !

SCORE	/33
mins	secs
TARGETS	/33
mins	secs

$6 \times 7 =$

$6 \times 6 =$ $6 \times 3 =$

$8 \times 6 =$ $6 \times 1 =$ $6 \times 5 =$

$0 \times 6 =$ $6 \times 9 =$ $2 \times 6 =$ $10 \times 6 =$

$9 \times 5 =$ $5 \times 4 =$ $5 \times 5 =$ $1 \times 5 =$ $4 \times 6 =$

$5 \times 6 =$ $5 \times 0 =$ $5 \times 8 =$ $3 \times 5 =$

$4 \times 7 =$ $4 \times 4 =$ $5 \times 10 =$ $5 \times 2 =$ $7 \times 5 =$

$6 \times 4 =$ $4 \times 3 =$ $8 \times 4 =$ $4 \times 1 =$

$2 \times 4 =$ $10 \times 4 =$ $0 \times 4 =$ $4 \times 9 =$ $4 \times 5 =$

NAME: _____

CLIMB THE WALL

4b

KEY FOCUS: *Multiplying by 4, 5 and 6*

How many can you get right ?

How fast can you do it ?

What will you try to get next time ?

YOU CAN DO IT !

SCORE	/33
mins	secs
TARGETS	/33
mins	secs

$9 \times 6 =$

$3 \times 6 =$ $5 \times 9 =$

$6 \times 5 =$ $4 \times 0 =$ $7 \times 6 =$

$6 \times 6 =$ $5 \times 4 =$ $4 \times 8 =$ $1 \times 6 =$

$10 \times 5 =$ $6 \times 2 =$ $7 \times 4 =$ $5 \times 1 =$ $4 \times 6 =$

$6 \times 4 =$ $4 \times 2 =$ $9 \times 4 =$ $5 \times 5 =$

$4 \times 5 =$ $5 \times 3 =$ $8 \times 5 =$ $6 \times 0 =$ $4 \times 10 =$

$5 \times 7 =$ $1 \times 4 =$ $6 \times 8 =$ $4 \times 4 =$

$0 \times 5 =$ $5 \times 6 =$ $3 \times 4 =$ $2 \times 5 =$ $6 \times 10 =$

NAME: _____

CLIMB THE WALL 4c

KEY FOCUS: *Multiplying by 4, 5 and 6*

How many can you get right ?

How fast can you do it ?

What will you try to get next time ?

YOU CAN DO IT !

SCORE	/33
mins	secs
TARGETS	/33
mins	secs

8 x 6 =

5 x 0 =   6 x 7 =

6 x 9 =   4 x 4 =   1 x 5 =

5 x 5 =   0 x 4 =   10 x 6 =   5 x 4 =

5 x 8 =   4 x 3 =   6 x 4 =   9 x 5 =   4 x 1 =

10 x 4 =   2 x 6 =   7 x 5 =   4 x 6 =

4 x 5 =   6 x 1 =   4 x 9 =   5 x 2 =   5 x 6 =

6 x 3 =   8 x 4 =   0 x 6 =   6 x 6 =

6 x 5 =   4 x 7 =   3 x 5 =   5 x 10 =   2 x 4 =

NAME: _____

CLIMB THE WALL

5a

KEY FOCUS: *Multiplying by 3, 6 and 9*

How many can you get right ?

How fast can you do it ?

What will you try to get next time ?

YOU CAN DO IT !

SCORE	/33
mins	secs
TARGETS	/33
mins	secs

6 x 9 =

2 x 9 =   9 x 9 =

7 x 9 =   4 x 9 =   0 x 9 =

9 x 5 =   9 x 1 =   8 x 9 =   9 x 3 =

6 x 8 =   3 x 6 =   9 x 6 =   6 x 6 =   10 x 9 =

6 x 2 =   7 x 6 =   6 x 0 =   5 x 6 =

3 x 9 =   3 x 3 =   6 x 4 =   6 x 10 =   1 x 6 =

10 x 3 =   2 x 3 =   3 x 7 =   3 x 1 =

3 x 5 =   0 x 3 =   8 x 3 =   4 x 3 =   6 x 3 =

NAME: _____

CLIMB THE WALL 5b

KEY FOCUS: *Multiplying by 3, 6 and 9*

How many can you get right ?

How fast can you do it ?

What will you try to get next time ?

YOU CAN DO IT !

SCORE	/33
mins	secs
TARGETS	/33
mins	secs

$9 \times 6 =$

$6 \times 6 =$ $5 \times 3 =$

$4 \times 6 =$ $9 \times 2 =$ $9 \times 10 =$

$7 \times 9 =$ $3 \times 0 =$ $9 \times 8 =$ $6 \times 7 =$

$6 \times 1 =$ $3 \times 9 =$ $10 \times 6 =$ $3 \times 3 =$ $9 \times 4 =$

$3 \times 6 =$ $6 \times 5 =$ $1 \times 9 =$ $3 \times 10 =$

$6 \times 9 =$ $3 \times 2 =$ $7 \times 3 =$ $0 \times 6 =$ $9 \times 3 =$

$8 \times 6 =$ $1 \times 3 =$ $9 \times 9 =$ $6 \times 3 =$

$3 \times 4 =$ $9 \times 0 =$ $5 \times 9 =$ $2 \times 6 =$ $3 \times 8 =$

NAME: _____

CLIMB THE WALL

5c

KEY FOCUS: *Multiplying by 3, 6 and 9*

How many can you get right ?

How fast can you do it ?

What will you try to get next time ?

YOU CAN DO IT !

SCORE	/33
mins	secs
TARGETS	/33
mins	secs

$9 \times 9 =$

$7 \times 6 =$ $3 \times 3 =$

$0 \times 3 =$ $6 \times 8 =$ $4 \times 9 =$

$6 \times 4 =$ $10 \times 9 =$ $9 \times 1 =$ $3 \times 5 =$

$0 \times 9 =$ $6 \times 3 =$ $1 \times 6 =$ $9 \times 6 =$ $4 \times 3 =$

$6 \times 9 =$ $8 \times 3 =$ $6 \times 0 =$ $9 \times 3 =$

$6 \times 2 =$ $3 \times 7 =$ $9 \times 5 =$ $6 \times 10 =$ $2 \times 9 =$

$10 \times 3 =$ $3 \times 6 =$ $8 \times 9 =$ $3 \times 1 =$

$6 \times 6 =$ $2 \times 3 =$ $3 \times 9 =$ $5 \times 6 =$ $9 \times 7 =$

CLIMB THE WALL 6a

NAME: _____

KEY FOCUS: *Multiplying by 4, 7 and 8*

How many can you get right ?

How fast can you do it ?

What will you try to get next time ?

YOU CAN DO IT !

SCORE	/33
mins	secs
TARGETS	/33
mins	secs

$8 \times 5 =$

$8 \times 8 =$ $8 \times 3 =$

$2 \times 8 =$ $8 \times 7 =$ $0 \times 8 =$

$6 \times 8 =$ $8 \times 1 =$ $8 \times 9 =$ $4 \times 8 =$

$5 \times 7 =$ $9 \times 7 =$ $7 \times 2 =$ $7 \times 7 =$ $10 \times 8 =$

$7 \times 6 =$ $7 \times 0 =$ $7 \times 8 =$ $3 \times 7 =$

$4 \times 9 =$ $6 \times 4 =$ $7 \times 10 =$ $1 \times 7 =$ $7 \times 4 =$

$8 \times 4 =$ $2 \times 4 =$ $4 \times 5 =$ $0 \times 4 =$

$4 \times 4 =$ $4 \times 1 =$ $4 \times 7 =$ $4 \times 3 =$ $10 \times 4 =$

NAME: _____

CLIMB THE WALL

6b

KEY FOCUS: *Multiplying by 4, 7 and 8*

How many can you get right ?

How fast can you do it ?

What will you try to get next time ?

YOU CAN DO IT !

SCORE	/ 33
mins	secs
TARGETS	/ 33
mins	secs

6 x 7 =

8 x 8 =   9 x 4 =

4 x 4 =   8 x 10 =   8 x 7 =

2 x 7 =   3 x 8 =   4 x 6 =   9 x 8 =

7 x 9 =   4 x 8 =   7 x 5 =   4 x 2 =   8 x 0 =

10 x 7 =   7 x 4 =   5 x 8 =   7 x 1 =

8 x 2 =   4 x 0 =   7 x 8 =   5 x 4 =   7 x 3 =

8 x 4 =   3 x 4 =   1 x 8 =   4 x 10 =

7 x 7 =   0 x 7 =   8 x 6 =   1 x 4 =   4 x 7 =

NAME: _____

CLIMB THE WALL

6c

KEY FOCUS: *Multiplying by 4, 7 and 8*

How many can you get right ?

How fast can you do it ?

What will you try to get next time ?

YOU CAN DO IT !

SCORE	/33
mins	secs
TARGETS	/33
mins	secs

$9 \times 7 =$

$8 \times 9 =$ $4 \times 7 =$

$10 \times 4 =$ $0 \times 8 =$ $7 \times 6 =$

$7 \times 8 =$ $4 \times 9 =$ $8 \times 8 =$ $2 \times 4 =$

$4 \times 3 =$ $2 \times 8 =$ $6 \times 4 =$ $7 \times 2 =$ $8 \times 5 =$

$10 \times 8 =$ $0 \times 4 =$ $7 \times 7 =$ $8 \times 4 =$

$8 \times 1 =$ $3 \times 7 =$ $4 \times 5 =$ $8 \times 7 =$ $4 \times 1 =$

$1 \times 7 =$ $4 \times 8 =$ $7 \times 0 =$ $5 \times 7 =$

$6 \times 8 =$ $7 \times 10 =$ $4 \times 4 =$ $8 \times 3 =$ $7 \times 4 =$

CLIMB THE WALL

7a

KEY FOCUS: *Multiplying by 6, 7 and 8*

How many can you get right ?

How fast can you do it ?

What will you try to get next time ?

YOU CAN DO IT !

SCORE	/33
mins	secs
TARGETS	/33
mins	secs

$8 \times 7 =$

$4 \times 8 =$

$8 \times 9 =$

$8 \times 8 =$

$2 \times 8 =$

$10 \times 8 =$

$8 \times 3 =$

$0 \times 8 =$

$6 \times 8 =$

$8 \times 1 =$

$1 \times 7 =$

$7 \times 6 =$

$7 \times 4 =$

$7 \times 8 =$

$8 \times 5 =$

$9 \times 7 =$

$3 \times 7 =$

$7 \times 7 =$

$7 \times 0 =$

$6 \times 3 =$

$6 \times 7 =$

$7 \times 10 =$

$7 \times 2 =$

$5 \times 7 =$

$6 \times 6 =$

$4 \times 6 =$

$6 \times 9 =$

$6 \times 1 =$

$0 \times 6 =$

$8 \times 6 =$

$2 \times 6 =$

$10 \times 6 =$

$6 \times 5 =$

NAME: _____

CLIMB THE WALL 7b

KEY FOCUS: *Multiplying by 6, 7 and 8*

How many can you get right ?

How fast can you do it ?

What will you try to get next time ?

YOU CAN DO IT !

SCORE	/33
mins	secs
TARGETS	/33
mins	secs

$8 \times 6 =$

$7 \times 5 =$ $3 \times 8 =$

$9 \times 8 =$ $7 \times 7 =$ $6 \times 8 =$

$3 \times 6 =$ $8 \times 7 =$ $6 \times 4 =$ $8 \times 10 =$

$10 \times 7 =$ $9 \times 6 =$ $7 \times 3 =$ $6 \times 0 =$ $5 \times 8 =$

$8 \times 2 =$ $5 \times 6 =$ $6 \times 7 =$ $7 \times 1 =$

$0 \times 7 =$ $1 \times 6 =$ $4 \times 7 =$ $6 \times 10 =$ $8 \times 8 =$

$2 \times 7 =$ $7 \times 8 =$ $6 \times 6 =$ $1 \times 8 =$

$8 \times 4 =$ $7 \times 6 =$ $8 \times 0 =$ $6 \times 2 =$ $7 \times 9 =$

CLIMB THE WALL 7c

KEY FOCUS: *Multiplying by 6, 7 and 8*

How many can you get right ?

How fast can you do it ?

What will you try to get next time ?

YOU CAN DO IT !

SCORE	/33
mins	secs
TARGETS	/33
mins	secs

$7 \times 7 =$

$8 \times 6 =$ $9 \times 7 =$

$7 \times 2 =$ $8 \times 9 =$ $6 \times 6 =$

$8 \times 1 =$ $7 \times 0 =$ $6 \times 8 =$ $4 \times 6 =$

$2 \times 6 =$ $10 \times 8 =$ $7 \times 6 =$ $6 \times 9 =$ $8 \times 3 =$

$7 \times 8 =$ $0 \times 6 =$ $8 \times 5 =$ $3 \times 7 =$

$1 \times 7 =$ $4 \times 8 =$ $6 \times 7 =$ $7 \times 10 =$ $6 \times 3 =$

$7 \times 4 =$ $6 \times 1 =$ $8 \times 7 =$ $6 \times 5 =$

$8 \times 8 =$ $5 \times 7 =$ $10 \times 6 =$ $2 \times 8 =$ $0 \times 8 =$

NAME: _____

CLIMB THE WALL

KEY FOCUS: *Multiplying by 7, 8 and 9*

8a

How many can you get right ?

How fast can you do it ?

What will you try to get next time ?

YOU CAN DO IT !

SCORE	/ 33
mins	secs
TARGETS	/ 33
mins	secs

$9 \times 7 =$

$9 \times 5 =$ $8 \times 9 =$

$0 \times 9 =$ $9 \times 3 =$ $6 \times 9 =$

$9 \times 9 =$ $4 \times 9 =$ $9 \times 1 =$ $10 \times 9 =$

$8 \times 4 =$ $5 \times 8 =$ $8 \times 2 =$ $7 \times 8 =$ $2 \times 9 =$

$1 \times 8 =$ $8 \times 6 =$ $8 \times 0 =$ $8 \times 10 =$

$7 \times 9 =$ $6 \times 7 =$ $9 \times 8 =$ $3 \times 8 =$ $8 \times 8 =$

$0 \times 7 =$ $4 \times 7 =$ $7 \times 7 =$ $7 \times 1 =$

$7 \times 5 =$ $2 \times 7 =$ $10 \times 7 =$ $7 \times 3 =$ $8 \times 7 =$

CLIMB THE WALL

8b

KEY FOCUS: *Multiplying by 7, 8 and 9*

How many can you get right ?

How fast can you do it ?

What will you try to get next time ?

YOU CAN DO IT !

SCORE	/33
mins	secs
TARGETS	/33
mins	secs

$8 \times 8 =$

$7 \times 6 =$ $3 \times 9 =$

$9 \times 7 =$ $4 \times 8 =$ $8 \times 5 =$

$5 \times 9 =$ $7 \times 0 =$ $8 \times 7 =$ $3 \times 7 =$

$8 \times 9 =$ $1 \times 9 =$ $7 \times 7 =$ $2 \times 8 =$ $9 \times 6 =$

$8 \times 3 =$ $9 \times 9 =$ $0 \times 8 =$ $7 \times 10 =$

$7 \times 9 =$ $1 \times 7 =$ $10 \times 8 =$ $9 \times 2 =$ $7 \times 8 =$

$5 \times 7 =$ $9 \times 0 =$ $7 \times 4 =$ $9 \times 8 =$

$6 \times 8 =$ $7 \times 2 =$ $9 \times 4 =$ $8 \times 1 =$ $9 \times 10 =$

CLIMB THE WALL

8c

NAME: _____

KEY FOCUS: *Multiplying by 7, 8 and 9*

How many can you get right ?

How fast can you do it ?

What will you try to get next time ?

YOU CAN DO IT !

SCORE	/33
mins	secs
TARGETS	/33
mins	secs

$9 \times 8 =$

$6 \times 9 =$　　$7 \times 8 =$

$9 \times 5 =$　　$7 \times 4 =$　　$8 \times 0 =$

$10 \times 7 =$　　$7 \times 9 =$　　$8 \times 8 =$　　$1 \times 8 =$

$4 \times 8 =$　　$7 \times 7 =$　　$5 \times 8 =$　　$9 \times 9 =$　　$8 \times 2 =$

$8 \times 6 =$　　$7 \times 3 =$　　$8 \times 10 =$　　$0 \times 9 =$

$2 \times 7 =$　　$8 \times 9 =$　　$7 \times 1 =$　　$9 \times 3 =$　　$3 \times 8 =$

$6 \times 7 =$　　$4 \times 9 =$　　$8 \times 7 =$　　$9 \times 1 =$

$7 \times 5 =$　　$10 \times 9 =$　　$0 \times 7 =$　　$9 \times 7 =$　　$2 \times 9 =$

NAME: _____

CLIMB THE WALL 9a

KEY FOCUS: *Multiplying by 6, 7, 8 and 9*

How many can you get right ?

How fast can you do it ?

What will you try to get next time ?

YOU CAN DO IT !

SCORE	/33
mins	secs
TARGETS	/33
mins	secs

$9 \times 7 =$

$8 \times 8 =$ $5 \times 9 =$

$8 \times 3 =$ $6 \times 7 =$ $9 \times 6 =$

$9 \times 10 =$ $7 \times 2 =$ $9 \times 4 =$ $8 \times 6 =$

$9 \times 2 =$ $4 \times 6 =$ $8 \times 9 =$ $7 \times 7 =$ $10 \times 6 =$

$7 \times 4 =$ $7 \times 9 =$ $6 \times 3 =$ $4 \times 8 =$

$6 \times 5 =$ $7 \times 8 =$ $9 \times 9 =$ $6 \times 9 =$ $7 \times 10 =$

$3 \times 7 =$ $10 \times 8 =$ $6 \times 6 =$ $3 \times 9 =$

$8 \times 5 =$ $2 \times 6 =$ $6 \times 8 =$ $5 \times 7 =$ $2 \times 8 =$

NAME: _____

CLIMB THE WALL

9b

KEY FOCUS: *Multiplying by 6, 7, 8 and 9*

How many can you get right ?

How fast can you do it ?

What will you try to get next time ?

YOU CAN DO IT !

SCORE	/33
mins	secs
TARGETS	/33
mins	secs

$9 \times 7 =$

$7 \times 8 =$ $8 \times 10 =$

$6 \times 7 =$ $9 \times 5 =$ $8 \times 4 =$

$9 \times 3 =$ $8 \times 8 =$ $6 \times 9 =$ $9 \times 8 =$

$7 \times 5 =$ $10 \times 7 =$ $3 \times 8 =$ $7 \times 7 =$ $4 \times 9 =$

$8 \times 9 =$ $6 \times 6 =$ $4 \times 7 =$ $10 \times 9 =$

$6 \times 8 =$ $3 \times 6 =$ $9 \times 6 =$ $5 \times 8 =$ $7 \times 6 =$

$9 \times 9 =$ $6 \times 4 =$ $8 \times 7 =$ $7 \times 3 =$

$5 \times 6 =$ $2 \times 9 =$ $6 \times 10 =$ $8 \times 6 =$ $7 \times 9 =$

NAME: _____

CLIMB THE WALL

9c

KEY FOCUS: *Multiplying by 6, 7, 8 and 9*

How many can you get right ?

How fast can you do it ?

What will you try to get next time ?

YOU CAN DO IT !

SCORE	33
mins	secs
TARGETS	33
mins	secs

8 x 8 =

6 x 7 =   8 x 9 =

7 x 8 =   6 x 9 =   8 x 3 =

6 x 8 =   3 x 9 =   10 x 8 =   7 x 9 =

8 x 5 =   9 x 7 =   5 x 7 =   6 x 6 =   9 x 2 =

9 x 6 =   7 x 10 =   6 x 3 =   9 x 8 =

7 x 4 =   8 x 6 =   3 x 7 =   9 x 10 =   4 x 8 =

4 x 6 =   9 x 9 =   6 x 5 =   8 x 7 =

5 x 9 =   7 x 7 =   10 x 6 =   9 x 4 =   7 x 6 =

NAME: _____

CLIMB THE WALL

9d

KEY FOCUS: *Multiplying by 6, 7, 8 and 9*

How many can you get right ?

How fast can you do it ?

What will you try to get next time ?

YOU CAN DO IT !

SCORE	/33
mins	secs
TARGETS	/33
mins	secs

$6 \times 9 =$

$7 \times 8 =$ $9 \times 5 =$

$9 \times 9 =$ $6 \times 7 =$ $7 \times 3 =$

$9 \times 7 =$ $6 \times 8 =$ $8 \times 8 =$ $6 \times 10 =$

$8 \times 10 =$ $4 \times 9 =$ $3 \times 6 =$ $5 \times 8 =$ $7 \times 9 =$

$8 \times 7 =$ $6 \times 6 =$ $8 \times 9 =$ $10 \times 7 =$

$7 \times 6 =$ $9 \times 6 =$ $9 \times 3 =$ $7 \times 5 =$ $8 \times 6 =$

$8 \times 4 =$ $9 \times 8 =$ $10 \times 9 =$ $6 \times 4 =$

$2 \times 9 =$ $7 \times 7 =$ $3 \times 8 =$ $4 \times 7 =$ $5 \times 6 =$

NAME: _____

CLIMB THE WALL

9e

KEY FOCUS: *Multiplying by 6, 7, 8 and 9*

How many can you get right ?

How fast can you do it ?

What will you try to get next time ?

YOU CAN DO IT !

SCORE	/33
mins	secs
TARGETS	/33
mins	secs

$4 \times 8 =$

$8 \times 8 =$ $9 \times 7 =$

$3 \times 7 =$ $9 \times 8 =$ $7 \times 6 =$

$7 \times 10 =$ $8 \times 5 =$ $6 \times 6 =$ $8 \times 7 =$

$4 \times 6 =$ $9 \times 9 =$ $6 \times 9 =$ $9 \times 4 =$ $6 \times 8 =$

$5 \times 7 =$ $8 \times 9 =$ $6 \times 7 =$ $9 \times 10 =$

$7 \times 7 =$ $8 \times 3 =$ $9 \times 2 =$ $7 \times 9 =$ $10 \times 6 =$

$6 \times 5 =$ $9 \times 6 =$ $7 \times 8 =$ $3 \times 9 =$

$8 \times 6 =$ $7 \times 4 =$ $10 \times 8 =$ $6 \times 3 =$ $5 \times 9 =$

Climb The Wall

ANSWERS

DATE	TASK	SCORE	TIME	TARGET(S)

ANSWERS
CLIMB THE WALL 1b

KEY FOCUS: Multiplying by 2, 5 and 10

What score/time did you get?
Are you pleased with it?
What could you do to improve your score or time?
What work do you need to do now?

$8 \times 5 = 40$

$3 \times 10 = 30$ $5 \times 7 = 35$

$10 \times 4 = 40$ $2 \times 9 = 18$ $5 \times 0 = 0$

$2 \times 8 = 16$ $4 \times 2 = 8$ $10 \times 6 = 60$ $4 \times 5 = 20$

$5 \times 10 = 50$ $2 \times 0 = 0$ $5 \times 3 = 15$ $9 \times 5 = 45$ $1 \times 10 = 10$

$2 \times 10 = 20$ $6 \times 2 = 12$ $2 \times 5 = 10$ $10 \times 10 = 100$

$5 \times 5 = 25$ $2 \times 7 = 14$ $7 \times 10 = 70$ $5 \times 2 = 10$

$5 \times 1 = 5$ $8 \times 10 = 80$ $2 \times 3 = 6$ $0 \times 10 = 0$

$10 \times 5 = 50$ $2 \times 2 = 4$ $9 \times 10 = 90$ $6 \times 5 = 30$ $10 \times 2 = 20$

ANSWERS
CLIMB THE WALL 1a

KEY FOCUS: Multiplying by 2, 5 and 10

What score/time did you get?
Are you pleased with it?
What could you do to improve your score or time?
What work do you need to do now?

$10 \times 10 = 100$

$10 \times 8 = 80$ $7 \times 10 = 70$

$5 \times 10 = 50$ $10 \times 0 = 0$ $3 \times 10 = 30$

$1 \times 10 = 10$ $10 \times 6 = 60$ $9 \times 10 = 90$ $10 \times 2 = 20$

$5 \times 4 = 20$ $0 \times 5 = 0$ $2 \times 5 = 10$ $10 \times 4 = 40$

$5 \times 6 = 30$ $3 \times 5 = 15$ $5 \times 10 = 50$ $1 \times 5 = 5$

$3 \times 2 = 6$ $5 \times 5 = 25$ $5 \times 8 = 40$ $7 \times 5 = 35$

$2 \times 10 = 20$ $2 \times 2 = 4$ $7 \times 2 = 14$ $2 \times 6 = 12$

$9 \times 2 = 18$ $1 \times 2 = 2$ $2 \times 8 = 16$ $0 \times 2 = 0$ $2 \times 4 = 8$

$9 \times 5 = 45$ $5 \times 2 = 10$

ANSWERS
CLIMB THE WALL

2a

KEY FOCUS: *Multiplying by 2, 3 and 4*

What score/time did you get?
Are you pleased with it?
What could you do to improve your score or time?
What work do you need to do now?

4 x 4 = 16

10 x 4 = 40   4 x 3 = 12

0 x 4 = 0   4 x 5 = 20   8 x 4 = 32

2 x 4 = 8   4 x 7 = 28   4 x 1 = 4   4 x 9 = 36

3 x 3 = 9   9 x 3 = 27   3 x 4 = 12   3 x 0 = 0   6 x 4 = 24

3 x 6 = 18   1 x 3 = 3   3 x 8 = 24   3 x 2 = 6

2 x 7 = 14   4 x 2 = 8   5 x 3 = 15   3 x 10 = 30   7 x 3 = 21

2 x 1 = 2   2 x 5 = 10   8 x 2 = 16   2 x 2 = 4

6 x 2 = 12   0 x 2 = 0   2 x 9 = 18   2 x 3 = 6   10 x 2 = 20

ANSWERS
CLIMB THE WALL

1c

KEY FOCUS: *Multiplying by 2, 5 and 10*

What score/time did you get?
Are you pleased with it?
What could you do to improve your score or time?
What work do you need to do now?

9 x 5 = 45

5 x 5 = 25   10 x 10 = 100

0 x 10 = 0   5 x 10 = 50   4 x 2 = 8

10 x 2 = 20   2 x 3 = 6   2 x 1 = 2   4 x 10 = 40

5 x 0 = 0   5 x 6 = 30   2 x 9 = 18   10 x 1 = 10

5 x 8 = 40   0 x 2 = 0   10 x 3 = 30   2 x 7 = 14

1 x 5 = 5   10 x 7 = 70   6 x 2 = 12   10 x 9 = 90   5 x 2 = 10

3 x 5 = 15   2 x 10 = 20   2 x 2 = 4   7 x 5 = 35

6 x 10 = 60   2 x 5 = 10   10 x 5 = 50   8 x 2 = 16   5 x 4 = 20

8 x 10 = 80

Worksheet 2c

ANSWERS

CLIMB THE WALL 2c

© 2014 Tony Colledge

KEY FOCUS: *Multiplying by 2, 3 and 4*

What score/time did you get ?
Are you pleased with it ?
What could you do to improve your score or time ?
What work do you need to do now ?

4 x 9 = 36
5 x 3 = 15   0 x 2 = 0
2 x 9 = 18   3 x 3 = 9   2 x 5 = 10
3 x 0 = 0   6 x 4 = 24   3 x 10 = 30   7 x 3 = 21
3 x 6 = 18   2 x 1 = 2   8 x 2 = 16   3 x 4 = 12   4 x 2 = 8
10 x 2 = 20   0 x 4 = 0   6 x 2 = 12   1 x 3 = 3
8 x 4 = 32   4 x 1 = 4   9 x 3 = 27   3 x 2 = 6
4 x 7 = 28   2 x 2 = 4   10 x 4 = 40   2 x 3 = 6
4 x 5 = 20   3 x 8 = 24   4 x 4 = 16   2 x 7 = 14   2 x 4 = 8

Worksheet 2b

ANSWERS

CLIMB THE WALL 2b

© 2014 Tony Colledge

KEY FOCUS: *Multiplying by 2, 3 and 4*

What score/time did you get ?
Are you pleased with it ?
What could you do to improve your score or time ?
What work do you need to do now ?

3 x 7 = 21
9 x 4 = 36
4 x 3 = 12   4 x 10 = 40   3 x 3 = 9
5 x 4 = 20   3 x 1 = 3   4 x 8 = 32   2 x 4 = 8
5 x 2 = 10   6 x 3 = 18   2 x 0 = 0
9 x 2 = 18   2 x 3 = 6   0 x 3 = 0   2 x 10 = 20   1 x 4 = 4   3 x 2 = 6
2 x 8 = 16   3 x 9 = 27   4 x 4 = 16   2 x 6 = 12
4 x 6 = 24   1 x 2 = 2   10 x 3 = 30   4 x 2 = 8   8 x 3 = 24
3 x 4 = 12   3 x 5 = 15   4 x 0 = 0
7 x 2 = 14   2 x 2 = 4   7 x 4 = 28

ANSWERS
CLIMB THE WALL

 **3b**

KEY FOCUS: *Multiplying by 3, 4 and 5*

What score/time did you get?
Are you pleased with it?
What could you do to improve your score or time?
What work do you need to do now?

$9 \times 5 = 45$

$3 \times 8 = 24$ $5 \times 5 = 25$

$4 \times 3 = 12$ $5 \times 4 = 20$ $4 \times 9 = 36$

$5 \times 3 = 15$ $3 \times 10 = 30$ $4 \times 1 = 4$ $7 \times 5 = 35$

$10 \times 4 = 40$ $4 \times 7 = 28$ $5 \times 0 = 0$ $4 \times 4 = 16$

$9 \times 3 = 27$ $0 \times 4 = 0$ $5 \times 6 = 30$ $3 \times 5 = 15$

$5 \times 2 = 10$ $3 \times 4 = 12$ $5 \times 10 = 50$ $3 \times 2 = 6$ $6 \times 4 = 24$

$3 \times 3 = 9$ $8 \times 4 = 32$ $1 \times 5 = 5$ $3 \times 6 = 18$

$7 \times 3 = 21$ $2 \times 4 = 8$ $4 \times 5 = 20$ $3 \times 0 = 0$ $5 \times 8 = 40$

ANSWERS
CLIMB THE WALL

 **3a**

KEY FOCUS: *Multiplying by 3, 4 and 5*

What score/time did you get?
Are you pleased with it?
What could you do to improve your score or time?
What work do you need to do now?

$10 \times 5 = 50$

$5 \times 5 = 25$ $2 \times 5 = 10$ $4 \times 5 = 20$

$0 \times 5 = 0$ $8 \times 4 = 32$ $5 \times 1 = 5$

$6 \times 5 = 30$ $5 \times 3 = 15$ $4 \times 4 = 16$ $5 \times 9 = 45$ $7 \times 4 = 28$ $5 \times 7 = 35$

$3 \times 4 = 12$ $4 \times 10 = 40$ $4 \times 8 = 32$ $4 \times 0 = 0$ $4 \times 6 = 24$

$3 \times 5 = 15$ $10 \times 3 = 30$ $9 \times 4 = 36$ $1 \times 4 = 4$ $5 \times 4 = 20$

$6 \times 3 = 18$ $2 \times 3 = 6$ $3 \times 7 = 21$ $3 \times 3 = 9$

$0 \times 3 = 0$ $4 \times 3 = 12$ $3 \times 9 = 27$ $3 \times 1 = 3$ $8 \times 3 = 24$

ANSWERS

CLIMB THE WALL 4a

KEY FOCUS: *Multiplying by 4, 5 and 6*

What score/time did you get ?
Are you pleased with it ?
What could you do to improve your score or time ?
What work do you need to do now ?

$6 \times 7 = 42$

$6 \times 6 = 36$ $6 \times 3 = 18$

$8 \times 6 = 48$ $6 \times 1 = 6$ $6 \times 5 = 30$

$0 \times 6 = 0$ $6 \times 9 = 54$ $2 \times 6 = 12$ $10 \times 6 = 60$

$9 \times 5 = 45$ $5 \times 4 = 20$ $5 \times 5 = 25$ $1 \times 5 = 5$ $4 \times 6 = 24$

$5 \times 6 = 30$ $5 \times 0 = 0$ $5 \times 8 = 40$ $3 \times 5 = 15$

$4 \times 7 = 28$ $4 \times 4 = 16$ $5 \times 10 = 50$ $5 \times 2 = 10$ $7 \times 5 = 35$

$6 \times 4 = 24$ $4 \times 3 = 12$ $8 \times 4 = 32$ $4 \times 1 = 4$

$2 \times 4 = 8$ $10 \times 4 = 40$ $0 \times 4 = 0$ $4 \times 9 = 36$ $4 \times 5 = 20$

ANSWERS

CLIMB THE WALL 3c

KEY FOCUS: *Multiplying by 3, 4 and 5*

What score/time did you get ?
Are you pleased with it ?
What could you do to improve your score or time ?
What work do you need to do now ?

$7 \times 4 = 28$

$8 \times 5 = 40$ $4 \times 4 = 16$

$5 \times 3 = 15$ $9 \times 4 = 36$ $0 \times 5 = 0$ $3 \times 4 = 12$

$3 \times 5 = 15$ $3 \times 9 = 27$ $2 \times 3 = 6$ $10 \times 5 = 50$ $3 \times 4 = 32$ $5 \times 1 = 5$

$0 \times 3 = 0$ $4 \times 10 = 40$ $3 \times 3 = 9$ $6 \times 5 = 30$

$4 \times 2 = 8$ $8 \times 3 = 24$ $3 \times 1 = 3$ $5 \times 7 = 35$ $4 \times 3 = 12$

$6 \times 3 = 18$ $5 \times 4 = 20$ $10 \times 3 = 30$ $4 \times 0 = 0$

$4 \times 5 = 20$ $5 \times 5 = 25$ $5 \times 9 = 45$ $3 \times 7 = 21$ $2 \times 5 = 10$ $4 \times 6 = 24$

ANSWERS

CLIMB THE WALL

4c

KEY FOCUS: *Multiplying by 4, 5 and 6*

What score/time did you get?
Are you pleased with it?
What could you do to improve your score or time?
What work do you need to do now?

- 8 x 6 = 48
- 5 x 0 = 0
- 6 x 7 = 42
- 6 x 9 = 54
- 1 x 5 = 5
- 4 x 4 = 16
- 5 x 5 = 25
- 0 x 4 = 0
- 10 x 6 = 60
- 5 x 4 = 20
- 5 x 8 = 40
- 4 x 3 = 12
- 6 x 4 = 24
- 9 x 5 = 45
- 4 x 1 = 4
- 10 x 4 = 40
- 2 x 6 = 12
- 7 x 5 = 35
- 4 x 6 = 24
- 4 x 5 = 20
- 6 x 1 = 6
- 4 x 9 = 36
- 5 x 2 = 10
- 5 x 6 = 30
- 6 x 3 = 18
- 8 x 4 = 32
- 0 x 6 = 0
- 6 x 6 = 36
- 6 x 5 = 30
- 4 x 7 = 28
- 3 x 5 = 15
- 5 x 10 = 50
- 2 x 4 = 8

ANSWERS

CLIMB THE WALL

4b

KEY FOCUS: *Multiplying by 4, 5 and 6*

What score/time did you get?
Are you pleased with it?
What could you do to improve your score or time?
What work do you need to do now?

- 9 x 6 = 54
- 3 x 6 = 18
- 5 x 9 = 45
- 4 x 0 = 0
- 7 x 6 = 42
- 4 x 8 = 32
- 1 x 6 = 6
- 6 x 5 = 30
- 5 x 4 = 20
- 7 x 4 = 28
- 5 x 1 = 5
- 4 x 6 = 24
- 6 x 6 = 36
- 6 x 2 = 12
- 9 x 4 = 36
- 5 x 5 = 25
- 4 x 10 = 40
- 10 x 5 = 50
- 4 x 2 = 8
- 8 x 5 = 40
- 6 x 0 = 0
- 4 x 4 = 16
- 6 x 10 = 60
- 6 x 4 = 24
- 5 x 3 = 15
- 1 x 4 = 4
- 6 x 8 = 48
- 4 x 5 = 20
- 5 x 7 = 35
- 5 x 6 = 30
- 3 x 4 = 12
- 2 x 5 = 10
- 0 x 5 = 0

ANSWERS
CLIMB THE WALL — 5b

© 2014 Tony Colledge **KEY FOCUS:** *Multiplying by 3, 6 and 9*

What score/time did you get ?
Are you pleased with it ?
What could you do to improve your score or time ?
What work do you need to do now ?

9 x 6 = 54

6 x 6 = 36 | 5 x 3 = 15

4 x 6 = 24 | 9 x 2 = 18 | 9 x 10 = 90

7 x 9 = 63 | 3 x 0 = 0 | 9 x 8 = 72 | 6 x 7 = 42

6 x 1 = 6 | 3 x 9 = 27 | 10 x 6 = 60 | 3 x 3 = 9 | 9 x 4 = 36

3 x 6 = 18 | 6 x 5 = 30 | 1 x 9 = 9 | 3 x 10 = 30

6 x 9 = 54 | 3 x 2 = 6 | 7 x 3 = 21 | 0 x 6 = 0 | 9 x 3 = 27

8 x 6 = 48 | 1 x 3 = 3 | 9 x 9 = 81 | 6 x 3 = 18

3 x 4 = 12 | 9 x 0 = 0 | 5 x 9 = 45 | 2 x 6 = 12 | 3 x 8 = 24

ANSWERS
CLIMB THE WALL — 5a

© 2014 Tony Colledge **KEY FOCUS:** *Multiplying by 3, 6 and 9*

What score/time did you get ?
Are you pleased with it ?
What could you do to improve your score or time ?
What work do you need to do now ?

6 x 9 = 54

2 x 9 = 18 | 9 x 9 = 81

7 x 9 = 63 | 4 x 9 = 36 | 0 x 9 = 0

9 x 5 = 45 | 9 x 1 = 9 | 8 x 9 = 72 | 9 x 3 = 27

6 x 8 = 48 | 3 x 6 = 18 | 9 x 6 = 54 | 6 x 6 = 36 | 10 x 9 = 90

6 x 2 = 12 | 7 x 6 = 42 | 6 x 0 = 0 | 5 x 6 = 30

3 x 9 = 27 | 3 x 3 = 9 | 6 x 4 = 24 | 6 x 10 = 60 | 1 x 6 = 6

10 x 3 = 30 | 2 x 3 = 6 | 3 x 7 = 21 | 3 x 1 = 3

3 x 5 = 15 | 0 x 3 = 0 | 8 x 3 = 24 | 4 x 3 = 12 | 6 x 3 = 18

ANSWERS
CLIMB THE WALL — 6a

KEY FOCUS: Multiplying by 4, 7 and 8

What score/time did you get?
Are you pleased with it?
What could you do to improve your score or time?
What work do you need to do now?

$8 \times 5 = 40$

$8 \times 8 = 64$ $8 \times 3 = 24$

$2 \times 8 = 16$ $8 \times 7 = 56$ $0 \times 8 = 0$

$6 \times 8 = 48$ $8 \times 1 = 8$ $8 \times 9 = 72$ $4 \times 8 = 32$

$5 \times 7 = 35$ $9 \times 7 = 63$ $7 \times 2 = 14$ $7 \times 7 = 49$ $10 \times 8 = 80$

$7 \times 6 = 42$ $7 \times 0 = 0$ $7 \times 8 = 56$ $3 \times 7 = 21$

$4 \times 9 = 36$ $6 \times 4 = 24$ $7 \times 10 = 70$ $1 \times 7 = 7$ $7 \times 4 = 28$

$8 \times 4 = 32$ $2 \times 4 = 8$ $4 \times 5 = 20$ $0 \times 4 = 0$

$4 \times 4 = 16$ $4 \times 1 = 4$ $4 \times 7 = 28$ $4 \times 3 = 12$ $10 \times 4 = 40$

ANSWERS
CLIMB THE WALL — 5c

KEY FOCUS: Multiplying by 3, 6 and 9

What score/time did you get?
Are you pleased with it?
What could you do to improve your score or time?
What work do you need to do now?

$9 \times 9 = 81$

$7 \times 6 = 42$ $3 \times 3 = 9$

$0 \times 3 = 0$ $6 \times 8 = 48$ $4 \times 9 = 36$

$10 \times 9 = 90$ $1 \times 6 = 6$ $9 \times 1 = 9$ $3 \times 5 = 15$

$6 \times 3 = 18$ $8 \times 3 = 24$ $9 \times 6 = 54$ $6 \times 0 = 0$ $9 \times 3 = 27$ $4 \times 3 = 12$

$6 \times 9 = 54$ $9 \times 5 = 45$ $6 \times 10 = 60$ $2 \times 9 = 18$

$6 \times 2 = 12$ $3 \times 7 = 21$ $3 \times 6 = 18$ $8 \times 9 = 72$ $3 \times 1 = 3$

$10 \times 3 = 30$ $3 \times 9 = 27$ $5 \times 6 = 30$ $9 \times 7 = 63$

$0 \times 9 = 0$ $6 \times 6 = 36$ $2 \times 3 = 6$

ANSWERS
CLIMB THE WALL

6c

KEY FOCUS: *Multiplying by 4, 7 and 8*

What score/time did you get?
Are you pleased with it?
What could you do to improve your score or time?
What work do you need to do now?

$9 \times 7 = 63$

$8 \times 9 = 72$ $4 \times 7 = 28$

$10 \times 4 = 40$ $0 \times 8 = 0$ $7 \times 6 = 42$

$7 \times 8 = 56$ $4 \times 9 = 36$ $8 \times 8 = 64$ $2 \times 4 = 8$

$4 \times 3 = 12$ $2 \times 8 = 16$ $6 \times 4 = 24$ $7 \times 2 = 14$ $8 \times 5 = 40$

$10 \times 8 = 80$ $0 \times 4 = 0$ $7 \times 7 = 49$ $8 \times 4 = 32$

$8 \times 1 = 8$ $3 \times 7 = 21$ $4 \times 5 = 20$ $8 \times 7 = 56$ $4 \times 1 = 4$

$1 \times 7 = 7$ $7 \times 0 = 0$ $4 \times 8 = 32$ $5 \times 7 = 35$

$6 \times 8 = 48$ $7 \times 10 = 70$ $4 \times 4 = 16$ $8 \times 3 = 24$ $7 \times 4 = 28$

ANSWERS
CLIMB THE WALL

6b

KEY FOCUS: *Multiplying by 4, 7 and 8*

What score/time did you get?
Are you pleased with it?
What could you do to improve your score or time?
What work do you need to do now?

$6 \times 7 = 42$

$8 \times 8 = 64$ $9 \times 4 = 36$

$4 \times 4 = 16$ $8 \times 10 = 80$ $8 \times 7 = 56$

$2 \times 7 = 14$ $3 \times 8 = 24$ $4 \times 6 = 24$ $9 \times 8 = 72$

$4 \times 8 = 32$ $7 \times 5 = 35$ $4 \times 2 = 8$ $8 \times 0 = 0$

$10 \times 7 = 70$ $7 \times 4 = 28$ $5 \times 8 = 40$ $7 \times 1 = 7$

$7 \times 9 = 63$ $4 \times 0 = 0$ $7 \times 8 = 56$ $5 \times 4 = 20$ $7 \times 3 = 21$

$8 \times 2 = 16$ $8 \times 4 = 32$ $3 \times 4 = 12$ $1 \times 8 = 8$ $4 \times 10 = 40$

$7 \times 7 = 49$ $0 \times 7 = 0$ $8 \times 6 = 48$ $1 \times 4 = 4$ $4 \times 7 = 28$

ANSWERS

CLIMB THE WALL

 7a

KEY FOCUS: *Multiplying by 6, 7 and 8*

What score/time did you get?
Are you pleased with it?
What could you do to improve your score or time?
What work do you need to do now?

$8 \times 7 = 56$

$4 \times 8 = 32$ $8 \times 9 = 72$

$8 \times 8 = 64$ $2 \times 8 = 16$ $10 \times 8 = 80$

$8 \times 3 = 24$ $0 \times 8 = 0$ $6 \times 8 = 48$ $8 \times 1 = 8$

$1 \times 7 = 7$ $7 \times 4 = 28$ $7 \times 8 = 56$ $8 \times 5 = 40$

$9 \times 7 = 63$ $3 \times 7 = 21$ $7 \times 7 = 49$ $7 \times 0 = 0$

$6 \times 7 = 42$ $7 \times 10 = 70$ $7 \times 2 = 14$ $5 \times 7 = 35$

$6 \times 3 = 18$ $6 \times 6 = 36$ $4 \times 6 = 24$ $6 \times 9 = 54$ $6 \times 1 = 6$

$0 \times 6 = 0$ $8 \times 6 = 48$ $2 \times 6 = 12$ $10 \times 6 = 60$ $6 \times 5 = 30$

ANSWERS

CLIMB THE WALL

 7b

KEY FOCUS: *Multiplying by 6, 7 and 8*

What score/time did you get?
Are you pleased with it?
What could you do to improve your score or time?
What work do you need to do now?

$8 \times 6 = 48$

$7 \times 5 = 35$ $3 \times 8 = 24$

$9 \times 8 = 72$ $7 \times 7 = 49$ $6 \times 8 = 48$

$3 \times 6 = 18$ $8 \times 7 = 56$ $6 \times 4 = 24$ $8 \times 10 = 80$

$10 \times 7 = 70$ $9 \times 6 = 54$ $7 \times 3 = 21$ $6 \times 0 = 0$ $5 \times 8 = 40$

$8 \times 2 = 16$ $5 \times 6 = 30$ $6 \times 7 = 42$ $7 \times 1 = 7$

$1 \times 6 = 6$ $4 \times 7 = 28$ $6 \times 10 = 60$ $8 \times 8 = 64$

$0 \times 7 = 0$ $2 \times 7 = 14$ $7 \times 8 = 56$ $6 \times 6 = 36$ $1 \times 8 = 8$

$8 \times 4 = 32$ $7 \times 6 = 42$ $8 \times 0 = 0$ $6 \times 2 = 12$ $7 \times 9 = 63$

ANSWERS

CLIMB THE WALL

8a

KEY FOCUS: *Multiplying by 7, 8 and 9*

What score/time did you get ?
Are you pleased with it ?
What could you do to improve your score or time ?
What work do you need to do now ?

$9 \times 7 = 63$

$9 \times 5 = 45$ $8 \times 9 = 72$

$0 \times 9 = 0$ $9 \times 3 = 27$ $6 \times 9 = 54$

$9 \times 9 = 81$ $4 \times 9 = 36$ $9 \times 1 = 9$ $10 \times 9 = 90$

$8 \times 4 = 32$ $5 \times 8 = 40$ $8 \times 2 = 16$ $7 \times 8 = 56$ $2 \times 9 = 18$

$1 \times 8 = 8$ $8 \times 6 = 48$ $8 \times 0 = 0$ $8 \times 10 = 80$

$7 \times 9 = 63$ $6 \times 7 = 42$ $9 \times 8 = 72$ $3 \times 8 = 24$ $8 \times 8 = 64$

$0 \times 7 = 0$ $4 \times 7 = 28$ $7 \times 7 = 49$ $7 \times 1 = 7$

$7 \times 5 = 35$ $2 \times 7 = 14$ $10 \times 7 = 70$ $7 \times 3 = 21$ $8 \times 7 = 56$

ANSWERS

CLIMB THE WALL

7c

KEY FOCUS: *Multiplying by 6, 7 and 8*

What score/time did you get ?
Are you pleased with it ?
What could you do to improve your score or time ?
What work do you need to do now ?

$7 \times 7 = 49$

$8 \times 6 = 48$ $9 \times 7 = 63$

$7 \times 2 = 14$ $8 \times 9 = 72$ $6 \times 6 = 36$

$8 \times 1 = 8$ $6 \times 8 = 48$ $4 \times 6 = 24$

$10 \times 8 = 80$ $7 \times 6 = 42$ $6 \times 9 = 54$ $8 \times 3 = 24$

$7 \times 8 = 56$ $0 \times 6 = 0$ $8 \times 5 = 40$ $3 \times 7 = 21$

$2 \times 6 = 12$ $4 \times 8 = 32$ $6 \times 7 = 42$ $7 \times 10 = 70$ $6 \times 3 = 18$

$1 \times 7 = 7$ $6 \times 1 = 6$ $8 \times 7 = 56$ $6 \times 5 = 30$

$7 \times 4 = 28$ $10 \times 6 = 60$ $2 \times 8 = 16$ $0 \times 8 = 0$

$8 \times 8 = 64$ $5 \times 7 = 35$

8c

ANSWERS

CLIMB THE WALL

© 2014 Tony Colledge

KEY FOCUS: *Multiplying by 7, 8 and 9*

What score/time did you get ?
Are you pleased with it ?
What could you do to improve your score or time ?
What work do you need to do now ?

9 x 8 = 72

6 x 9 = 54   7 x 8 = 56

9 x 5 = 45   7 x 4 = 28   8 x 0 = 0

10 x 7 = 70   7 x 9 = 63   8 x 8 = 64   1 x 8 = 8

4 x 8 = 32   7 x 7 = 49   5 x 8 = 40   9 x 9 = 81   8 x 2 = 16

8 x 6 = 48   7 x 3 = 21   8 x 10 = 80   0 x 9 = 0

2 x 7 = 14   8 x 9 = 72   7 x 1 = 7   9 x 3 = 27   3 x 8 = 24

6 x 7 = 42   4 x 9 = 36   8 x 7 = 56   9 x 1 = 9

7 x 5 = 35   10 x 9 = 90   0 x 7 = 0   9 x 7 = 63   2 x 9 = 18

8b

ANSWERS

CLIMB THE WALL

© 2014 Tony Colledge

KEY FOCUS: *Multiplying by 7, 8 and 9*

What score/time did you get ?
Are you pleased with it ?
What could you do to improve your score or time ?
What work do you need to do now ?

8 x 8 = 64

7 x 6 = 42   3 x 9 = 27

9 x 7 = 63   4 x 8 = 32   8 x 5 = 40

5 x 9 = 45   7 x 0 = 0   8 x 7 = 56   3 x 7 = 21

8 x 3 = 24   7 x 7 = 49   2 x 8 = 16   9 x 6 = 54

9 x 9 = 81   0 x 8 = 0   7 x 10 = 70

7 x 9 = 63   1 x 7 = 7   10 x 8 = 80   9 x 2 = 18   7 x 8 = 56

5 x 7 = 35   9 x 0 = 0   7 x 4 = 28   9 x 8 = 72

6 x 8 = 48   7 x 2 = 14   9 x 4 = 36   8 x 1 = 8   9 x 10 = 90

ANSWERS

CLIMB THE WALL

9a

KEY FOCUS: *Multiplying by 6, 7, 8 and 9*

What score/time did you get ?

Are you pleased with it ?

What could you do to improve your score or time ?

What work do you need to do now ?

9 x 7 = 63

8 x 8 = 64   5 x 9 = 45

8 x 3 = 24   6 x 7 = 42   9 x 6 = 54

9 x 10 = 90   7 x 2 = 14   9 x 4 = 36   8 x 6 = 48

4 x 6 = 24   8 x 9 = 72   7 x 7 = 49   10 x 6 = 60

7 x 4 = 28   7 x 9 = 63   6 x 3 = 18   4 x 8 = 32

9 x 2 = 18   9 x 9 = 81   6 x 9 = 54   7 x 10 = 70

6 x 5 = 30   10 x 8 = 80   6 x 6 = 36   3 x 9 = 27

3 x 7 = 21   6 x 8 = 48   5 x 7 = 35   2 x 8 = 16

2 x 6 = 12

8 x 5 = 40

ANSWERS

CLIMB THE WALL

9b

KEY FOCUS: *Multiplying by 6, 7, 8 and 9*

What score/time did you get ?

Are you pleased with it ?

What could you do to improve your score or time ?

What work do you need to do now ?

9 x 7 = 63

7 x 8 = 56   8 x 10 = 80

6 x 7 = 42   9 x 5 = 45   8 x 4 = 32

8 x 8 = 64   6 x 9 = 54   4 x 9 = 36

9 x 3 = 27   10 x 7 = 70   3 x 8 = 24   7 x 7 = 49   9 x 8 = 72

8 x 9 = 72   6 x 6 = 36   4 x 7 = 28   10 x 9 = 90   7 x 6 = 42

3 x 6 = 18   9 x 6 = 54   5 x 8 = 40

9 x 9 = 81   6 x 4 = 24   8 x 7 = 56   7 x 3 = 21

2 x 9 = 18   6 x 10 = 60   8 x 6 = 48   7 x 9 = 63

6 x 8 = 48

5 x 6 = 30

ANSWERS — CLIMB THE WALL — 9d

© 2014 Tony Colledge

KEY FOCUS: *Multiplying by 6, 7, 8 and 9*

What score/time did you get?
Are you pleased with it?
What could you do to improve your score or time?
What work do you need to do now?

$6 \times 9 = 54$

$7 \times 8 = 56$ $9 \times 5 = 45$

$9 \times 9 = 81$ $6 \times 7 = 42$ $7 \times 3 = 21$

$9 \times 7 = 63$ $6 \times 8 = 48$ $8 \times 8 = 64$ $6 \times 10 = 60$

$8 \times 10 = 80$ $4 \times 9 = 36$ $3 \times 6 = 18$ $5 \times 8 = 40$ $7 \times 9 = 63$

$8 \times 7 = 56$ $6 \times 6 = 36$ $8 \times 9 = 72$ $10 \times 7 = 70$

$7 \times 6 = 42$ $9 \times 6 = 54$ $9 \times 3 = 27$ $7 \times 5 = 35$ $8 \times 6 = 48$

$8 \times 4 = 32$ $9 \times 8 = 72$ $10 \times 9 = 90$ $6 \times 4 = 24$

$2 \times 9 = 18$ $7 \times 7 = 49$ $3 \times 8 = 24$ $4 \times 7 = 28$ $5 \times 6 = 30$

ANSWERS — CLIMB THE WALL — 9c

© 2014 Tony Colledge

KEY FOCUS: *Multiplying by 6, 7, 8 and 9*

What score/time did you get?
Are you pleased with it?
What could you do to improve your score or time?
What work do you need to do now?

$8 \times 8 = 64$

$6 \times 7 = 42$ $8 \times 9 = 72$

$7 \times 8 = 56$ $6 \times 9 = 54$ $8 \times 3 = 24$

$6 \times 8 = 48$ $3 \times 9 = 27$ $10 \times 8 = 80$ $7 \times 9 = 63$

$8 \times 5 = 40$ $9 \times 7 = 63$ $5 \times 7 = 35$ $6 \times 6 = 36$ $9 \times 2 = 18$

$9 \times 6 = 54$ $7 \times 10 = 70$ $6 \times 3 = 18$ $9 \times 8 = 72$

$7 \times 4 = 28$ $8 \times 6 = 48$ $3 \times 7 = 21$ $9 \times 10 = 90$ $4 \times 8 = 32$

$4 \times 6 = 24$ $9 \times 9 = 81$ $6 \times 5 = 30$ $8 \times 7 = 56$

$5 \times 9 = 45$ $7 \times 7 = 49$ $10 \times 6 = 60$ $9 \times 4 = 36$ $7 \times 6 = 42$

ANSWERS
CLIMB THE WALL

9e

KEY FOCUS: *Multiplying by 6, 7, 8 and 9*

What score/time did you get ?
Are you pleased with it ?
What could you do to improve your score or time ?
What work do you need to do now ?

$4 \times 8 = 32$

$8 \times 8 = 64$ $9 \times 7 = 63$

$3 \times 7 = 21$ $9 \times 8 = 72$ $7 \times 6 = 42$

$7 \times 10 = 70$ $8 \times 5 = 40$ $6 \times 6 = 36$ $8 \times 7 = 56$

$9 \times 9 = 81$ $6 \times 9 = 54$ $9 \times 4 = 36$ $6 \times 8 = 48$

$4 \times 6 = 24$ $5 \times 7 = 35$ $8 \times 9 = 72$ $6 \times 7 = 42$ $9 \times 10 = 90$

$7 \times 7 = 49$ $8 \times 3 = 24$ $9 \times 2 = 18$ $7 \times 9 = 63$ $10 \times 6 = 60$

$6 \times 5 = 30$ $9 \times 6 = 54$ $7 \times 8 = 56$ $3 \times 9 = 27$

$8 \times 6 = 48$ $7 \times 4 = 28$ $10 \times 8 = 80$ $6 \times 3 = 18$ $5 \times 9 = 45$